Chemical Engineers in Medicine 2017

Topical Conference at the 2017 AIChE Annual Meeting

Minneapolis, Minnesota, USA
29 October – 3 November 2017

ISBN: 978-1-5108-5779-7

Printed from e-media with permission by:

Curran Associates, Inc.
57 Morehouse Lane
Red Hook, NY 12571

Printed by Curran Associates, Inc. (2018)

For permission requests, please contact AIChE
at the address below.

AIChE
120 Wall Street, FL 23
New York, NY 10005-4020

Phone: (800) 242-4363
Fax: (203) 775-5177

www.aiche.org

Additional copies of this publication are available from:

Curran Associates, Inc.
57 Morehouse Lane
Red Hook, NY 12571 USA
Phone: 845-758-0400
Fax: 845-758-2633
Email: curran@proceedings.com
Web: www.proceedings.com

TABLE OF CONTENTS

Some abstracts in this book refer to images that were not made available by the publisher

504784 Overcoming Biological Barriers for Drug Delivery

Abstract:

Effective delivery of drugs is a major problem in today's healthcare. At a fundamental level, the challenge of drug delivery reflects the fact that the drug distribution in the body is limited by body's natural metabolic processes and transport barriers. These biological barriers, while serving an important purpose of regulating body's metabolic functions, limit the drug dose that ultimately reaches the target site. Accordingly, many drugs fail to reach their full therapeutic potential. Our research aims at developing a fundamental understanding of body's key biological barriers such as skin and intestinal epithelium, and utilizing this understanding to develop novel means to negotiate these barriers to deliver drugs. I will present an overview of the lessons learned from our exploration of these biological barriers.

485869 Engineering the Vocal Cords

Abstract:

The sound of the human voice originates from the paired vocal cords in the larynx. Sound waves are generated when exhaled airflow from the lungs drives the cyclic collisions of the two vocal cords, hundreds of times per second. This oscillation is the fortunate outcome of tracheal pressure balanced with tissue elasticity, as air pressure fluctuations intermittently force opening and closing of the glottic valve. The vocal cord microstructure develops to permit and withstand this recurrent high-speed trauma from shear forces and impacts. Its layered structure and extracellular matrix (ECM) rich in elastic fibers and hyaluronic acid are unmatched by any other tissue.

Vocal cord injury disrupts this specialized ECM. As the structure stiffens, function follows form and the voice becomes harsh and effortful. Traditional repair efforts are disappointing, due to the ubiquitous challenges in remodeling scar tissue. Regenerative medicine and tissue engineering have been proposed as an alternative for these patients who suffer social isolation and occupational limitation due to their voice disability.

One investigational approach excises the vocal cord scar tissue and replaces it with a simple tissue-engineered structure of fibrin with adipose-derived multi-potent stromal cells. Tissue remodeling ensues, and the resulting vocal cords are able to vibrate for sound production in a rabbit model. Comparison with normal vocal cord transplantation shows superior function with the tissue-engineered structure. Results suggest that modulation of wound healing is required to preserve the complex ECM structure that permits vibratory function.

In this talk, the physiology and biomechanics of voice production will be reviewed. The unique vocal cord structure will be demonstrated in normal and diseased states. The status of cell-based therapies in the larynx will be reviewed, with special attention to the three-dimensional tissue-engineered implant developed in this laboratory. The ongoing challenge of elastic fiber production is defined.

504785 Re-engineering the tumor microenvironment to enhance efficacy of molecular and immune therapies in metastatic cancer

Abstract:

Pancreatic ductal adenocarcinoma (PDA) is an extremely lethal disease with the lowest 1-year and 5-year survival rates of any cancer. This is due, in part, to the extremely metastatic behavior of pancreas carcinoma cells and their extreme resistance to both chemical and radiotherapies. Importantly, we now know that a strong, but nevertheless unique, fibroinflammatory, or desmoplastic, response is present in the tumor microenvironments of PDA. Indeed, these features generate â€˜stromal resistanceâ€™ through physical barriers to effective treatment that include inordinately high interstitial fluid pressures, elevated tumor microenvironment stiffnesses, vascular collapse and low permeability; essentially creating a drug-free sanctuaries in PDA by critically limiting the delivery, diffusion, and convection of small molecule and immune therapeutics. To overcome these barriers, we develop strategies, that are informed by our work to define the fundamental physical behavior of solid tumors, to re-engineer the tumor stroma to normalize disease and enhance efficacy of therapeutic interventions. Indeed we have recently defined free and immobilized fluid phases within tumors and extracellular matrix components that drive elevated tumor pressures and hinder small molecule diffusion. By targeting these drivers of stromal resistance in the tumor, pressures are normalized, vascular patency is restored following re-expansion of microvasculature, and enhanced drug delivery and distribution results throughout the tumor mass from improved diffusion and convection. In the context of concomitant cytotoxic chemotherapy tumor volume is decreased and metastatic burden and overall survival can be significantly improved.

492370 Heat Transfer Effects on *Staphylococcus epidermidis* Biofilms: An *in vitro* Catheter Model

Abstract:

Bacterial biofilms are a leading cause of hospital-acquired infections. The development of biofilms on medical devices, such as hemodialysis catheters, can be understood as a consequence of adsorption, growth, and detachment; each effect is governed by self-assembly, fluid mechanics, and transport phenomena which are governed by fundamental chemical engineering principles. Here we show how the fundamental chemical engineering principles of fluid dynamics and heat transfer can be applied to prevent and/or remove bacterial biofilms in an *in vitro*, flow cell model of a dialysis catheter. Using *Staphylococcus epidermidis*, the most common bacterial species isolated from infected medical devices, we grow biofilms under physiologically relevant flow conditions and expose them to thermal degradation using a pre-warmed fluid that is injected into the model catheter. We establish that exposing biofilms to elevated temperatures changes the biofilm morphology and cell viability. Further, we discuss the mechanism by which this degradation occurs in light of the two parameters temperature elevation and exposure duration. Finally, we probe the resilient, adaptive nature of bacterial cells to investigate if they develop thermal resistance when exposed to the heat treatment. Understanding the response of these bacterial cells under thermal stress is a promising step toward the development of an *in situ* treatment/remediation method for biofilm growth in medical devices.

493184 Elucidating the Physiological Significance of Nitric Oxide Synthase (NOS) in *Staphylococcus aureus* through Metabolic Modeling

Abstract:

In recent years, whole genome sequencing has accelerated the reconstruction of genome scale metabolic models of several species. However, often these models do not accurately predict biological phenotypes due to inherent insufficiencies including metabolic gaps, unbalanced reactions and lack of key physiological information. In the present study, the genome-scale metabolic model of the bacterial pathogen, *Staphylococcus aureus* was reconstructed from ModelSEED and previous strain-specific models[1,2]. Reactions were elemental and charge balanced. Upon manual and automated gap-filling the model has 842 metabolic genes, 1430 metabolites and 1701 reactions including transport and export reactions. The model was applied to understanding the physiological function of *S. aureus* nitric oxide synthase (NOS).

The *S. aureus* NOS is an enzyme that converts arginine to citrulline and produces the highly reactive nitric oxide radical (NOÎ‡) as a byproduct. Under aerobic conditions, NOÎ‡ is further decomposed to nitrates and nitrites. Inactivation of *nos* decreases growth yields. However, the mechanism by which NOS activity enhances staphylococcal biomass production is unclear. Nuclear magnetic resonance (NMR) spectroscopic analysis of the *nos* mutant revealed decreased TCA cycle activity compared to its isogenic wild-type strain. It is conceivable that carbon-flux of arginine to 2-oxoglutarate through the glutamate node following NOS activity may be important for maintenance of TCA cycle activity. Alternately, our results also support a role for nitrites that indirectly result from NOS activity in stimulating terminal cytochrome (quinol oxidase) function and enhancing rate of respiration. The latter may also affect TCA cycle activity and biomass. Our analysis using the reconstructed genome scale metabolic model suggests that the latter possibility wherein nitrites affect respiration is more likely to modulate TCA cycle activity given that carbon flux through NOS is limited.

We have directly confirmed this prediction of the metabolic model by demonstrating that biomass production is rescued in the *nos* mutant following restoration of respiration by the addition of physiological concentrations of nitrite in the growth media. Finally, addition of 2-oxoglutarate (a TCA cycle metabolite) to the media does not restore biomass of the *nos* mutant to wild-type levels even though it is completely consumed by these strains. This suggests that carbon flux to the TCA cycle through the glutamate node is not a significant contributor of biomass in the *nos* mutant. These findings not only attest to the predictive power of our metabolic model but also highlight the biological role of bacterial NOS as an activator of quinol oxidase mediated aerobic respiration.

References:

1. Becker SA, Palsson BO. Genome-scale reconstruction of the metabolic network in Staphylococcus aureus N315: an initial draft to the two-dimensional annotation. *BMC microbiology*. Mar 07 2005;5:8.

2. Bosi E, Monk JM, Aziz RK, Fondi M, Nizet V, Palsson BO. Comparative genome-scale modelling of Staphylococcus aureus strains identifies strain-specific metabolic capabilities linked to pathogenicity. *Proc Natl Acad Sci U S A*. Jun 28 2016;113(26):E3801-3809.

493327 Regulating Fibrin Formation, Structure, and Mechanical Strength

Abstract:

Blood coagulation proceeds through a number of pathways within the coagulation cascade. These pathways can be initiated via exposure to negatively charged surfaces (intrinsic contact pathway) or through interaction with tissue factor (extrinsic pathway). However, the series of plasma protein reactions that occur to subsequently produce thrombin and fibrin can be drastically affected by a deficiency in zymogen factor XI (FXI), an integral protein in the coagulation cascade, which is activated by FXIIa and thrombin, or in zymogen factor XII, which is the starting point of the intrinsic pathway. Deficiencies in FXI result in prolonged bleeding times, suggesting that FXI plays a crucial role in regulating hemostatic plug formation. The structure of fibrin is integral to the kinetics and stability of thrombus formation, controlling clot strength and susceptibility to lysis. Thus, we used a suite of FXI inhibitors to measure the role of FXI on fibrin by analyzing the kinetics of fibrin formation, quantifying the extent of fibrin gel structure, and measuring the rheological properties of the transient coagulation. We show that FXI plays a key role in regulating the kinetics of fibrin formation and overall fibrin gel structure.

Roles of Conserved Tryptophans in Trimerization of HIV-1 Membrane-Proximal External Regions: Implications for Virucidal Design via Alchemical Free-Energy Molecular Simulations

Steven Gossert and Cameron Abrams
Drexel University
Department of Chemical and Biological Engineering

AIChE Annual Meeting
Minneapolis
Monday October 30th, 2017

HIV-1 entry mechanism occurs when Env binds to CD4 leading to the conformational cascade.

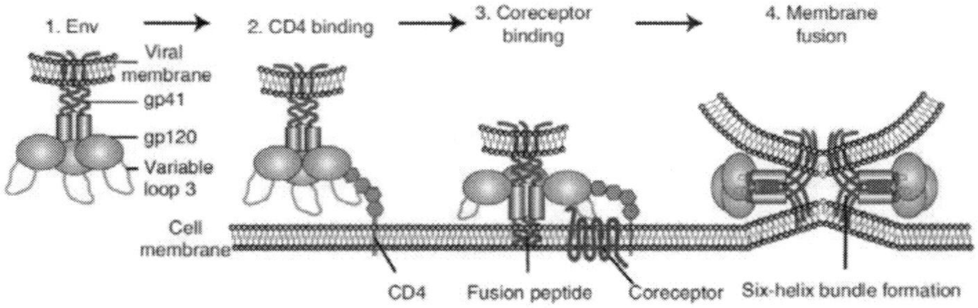

C.B. Wilen, J.C. Tilton, and R.W. Doms. HIV: Cell Binding and Entry. *Cold Spring Harbor Perspective in Medicine.* 2012.

2

DAVEI* acts on Env to trigger HIV-1 lysis shown through p24 release.

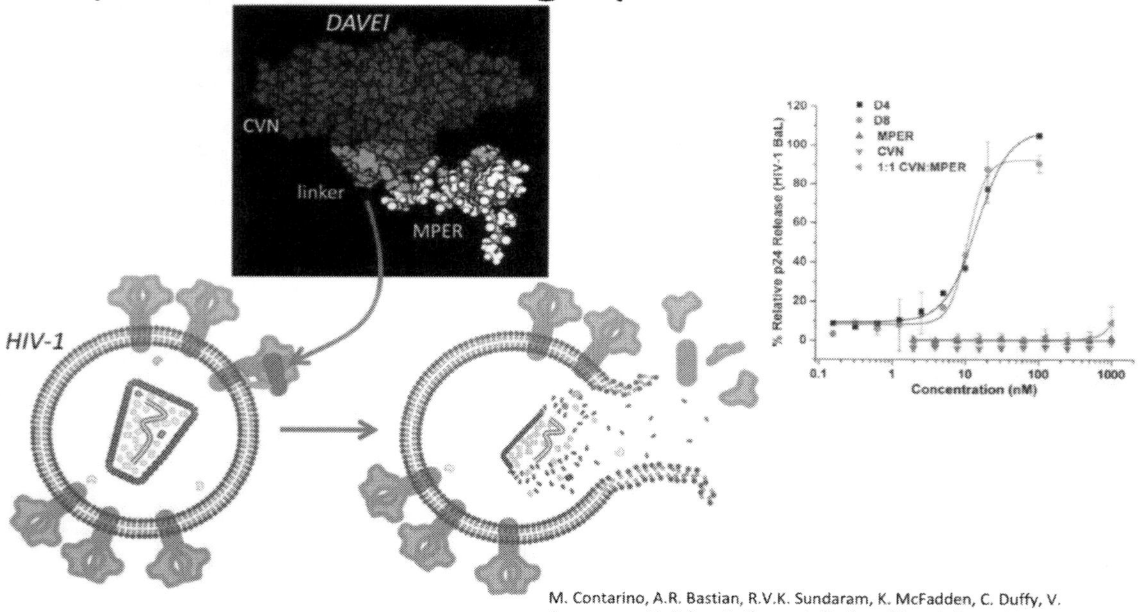

*Dual-Acting Virolytic Entry Inhibitor

M. Contarino, A.R. Bastian, R.V.K. Sundaram, K. McFadden, C. Duffy, V. Gangupomu, M. Baker, **C. Abrams**, I. Chaiken. Chimeric Cyanovirin-MPER Recombinantly Engineered Proteins Cause Cell-Free Virolysis of HIV-1. *Antimicrobial Agents and Chemotherapy*. 57: 4743-4750, 2013.

3

How does DAVEI MPER engage Env and cause viral lysis?

Hypothesis: DAVEI-MPER forms $MPER_3$ by disrupting the native Env-$MPER_3$ and replacing one of the native MPER causing the HIV-1 conformational cascade.

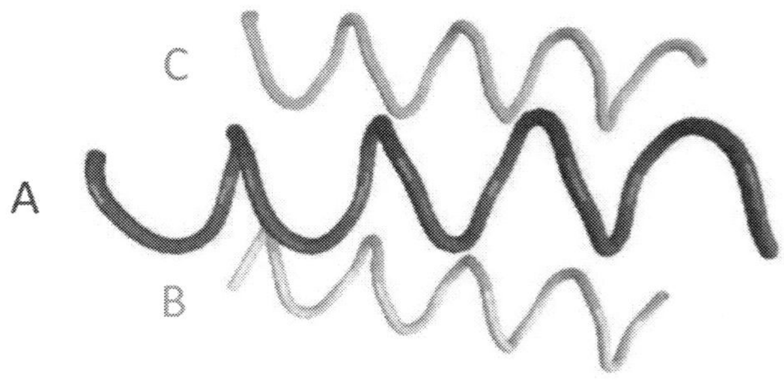

4

If MPER-MPER interactions are important then what role do tryptophan residues have in stability of the trimer?

Approach: Free Energy analysis on mutated $MPER_3$ structure to measure effect of mutation on trimer stability.

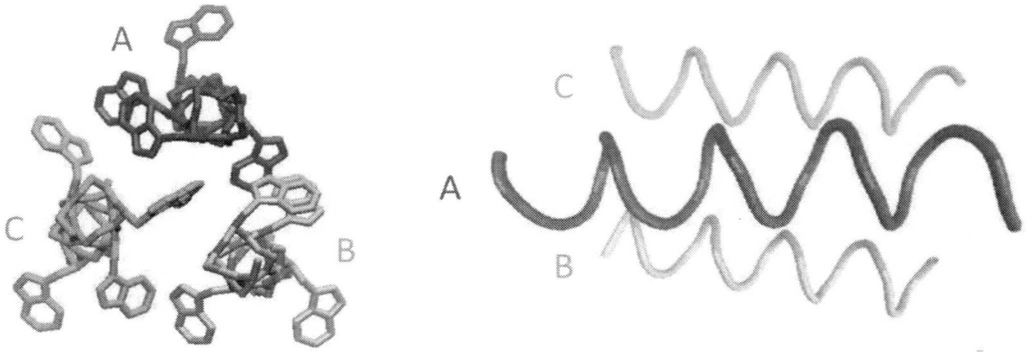

5

DAVEI approaches Env and CVN binds to surface glycans on gp120.

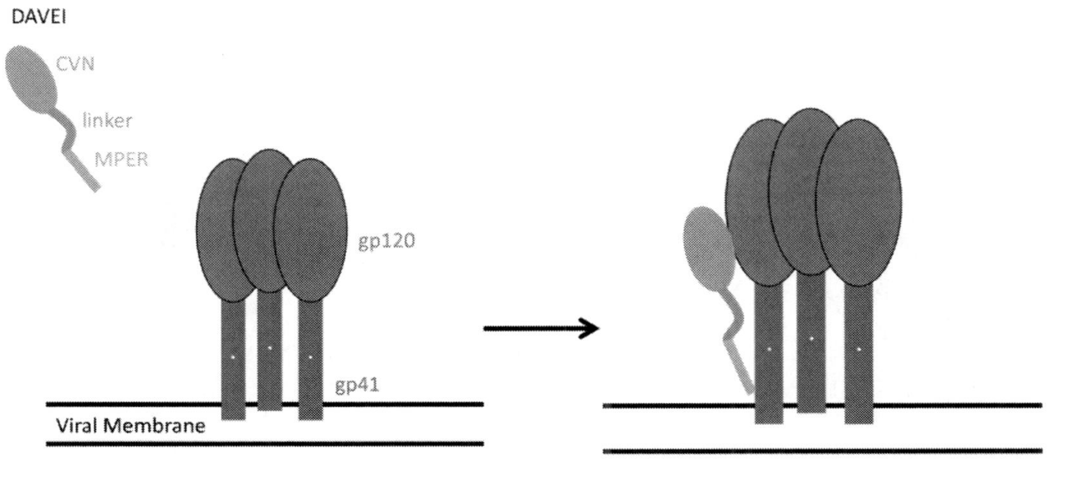

B. Parajuli, et al. Lytic Inactivation of Human Immunodeficiency Virus by Dual Engagement of gp120 and gp41 Domains in the Virus Env Protein Trimer, *Biochemistry. 2016*

M. Contarino, A.R. Bastian, R.V.K. Sundaram, K. McFadden, C. Duffy, V. Gangupomu, M. Baker, **C. Abrams**, I. Chaiken. Chimeric Cyanovirin-MPER Recombinantly Engineered Proteins Cause Cell-Free Virolysis of HIV-1. *Antimicrobial Agents and Chemotherapy*. 57: 4743-4750, 2013.

6

DAVEI-MPER approaches native Env $MPER_3$ on gp41.

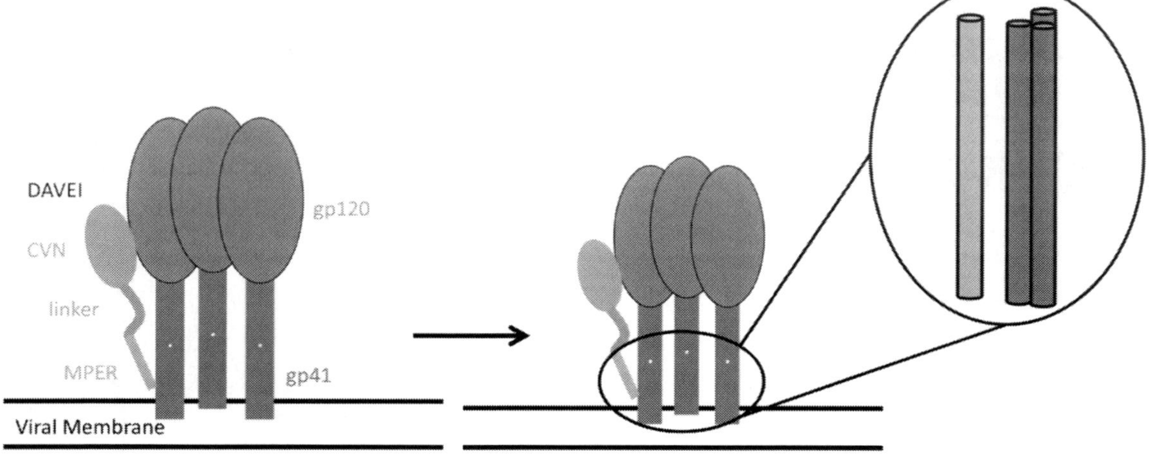

B. Parajuli, et al. Lytic Inactivation of Human Immunodeficiency Virus by Dual Engagement of gp120 and gp41 Domains in the Virus Env Protein Trimer, *Biochemistry. 2016*

M. Contarino, A.R. Bastian, R.V.K. Sundaram, K. McFadden, C. Duffy, V. Gangupomu, M. Baker, **C. Abrams**, I. Chaiken. Chimeric Cyanovirin-MPER Recombinantly Engineered Proteins Cause Cell-Free Virolysis of HIV-1. *Antimicrobial Agents and Chemotherapy*. 57: 4743-4750, 2013.

7

DAVEI-MPER replaces one Env MPER to create new $MPER_3$ that triggers conformational cascade of Env.

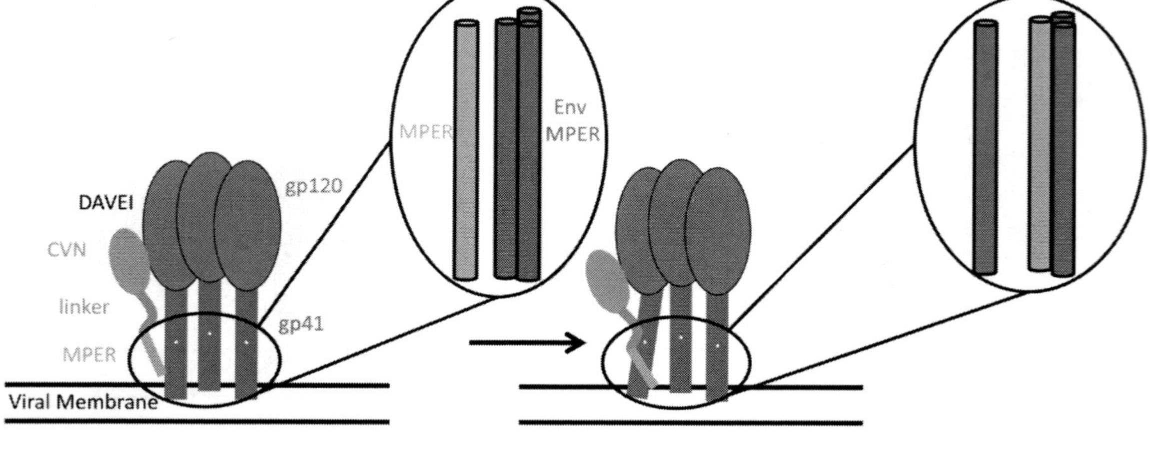

Hypothetical Model

8

Mutation Studies on DAVEI-MPER tryptophan determined only first tryptophan is essential in causing lysis.

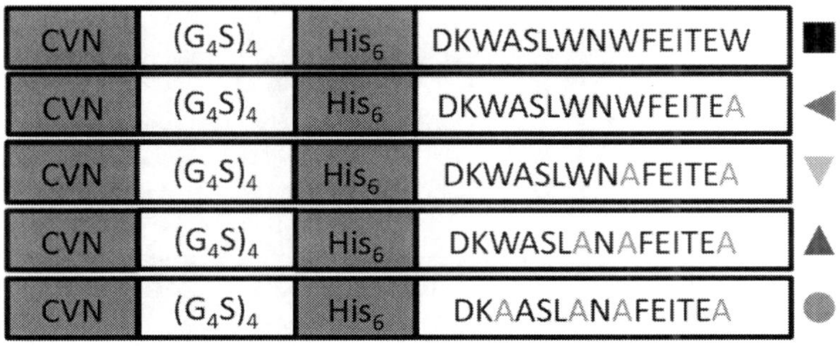

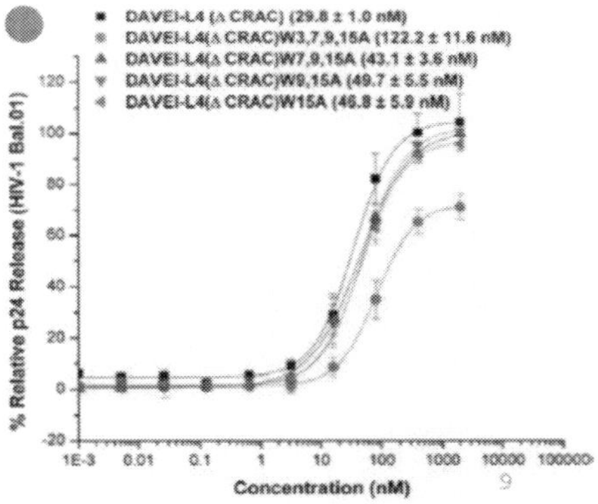

B. Parajuli, et al. Lytic Inactivation of Human Immunodeficiency Virus by Dual Engagement of gp120 and gp41 Domains in the Virus Env Protein Trimer, *Biochemistry. 2016*

9

MPER$_3$ was generated using the 3G9R crystal structure for free energy testing of the mutations.

Generated Sequence (HXBc2 numbering):

666 670 672 678

661-ELDKWASLWNWFNITNWLWYIK-684

CRAC sequence

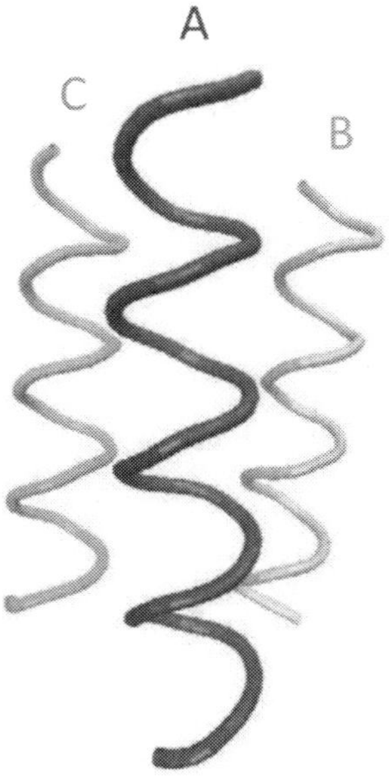

HXBc2 Sequence ID of mutated residues:
W666, W670, W672, W678

J. Liu, M. Lu. Structure of the HIV-1 gp41 Membrane-Proximal Ectodomain Region in a Putative Prefusion Conformation. *Biochemistry* 48:2915-2923 2009.c

10

Thermodynamic cycle of the mutations and trimerization of MPER where ΔG_3 and ΔG_4 can be calculated using Free Energy Perturbation.

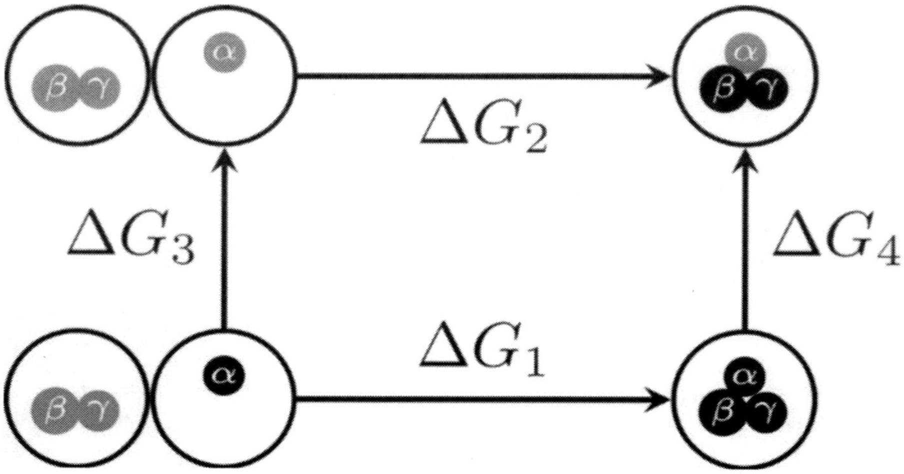

$$\Delta\Delta G \equiv \Delta G_2 - \Delta G_1 = \Delta G_4 - \Delta G_3$$

11

FEP generates a free energy profile along λ to calculate the ΔΔG

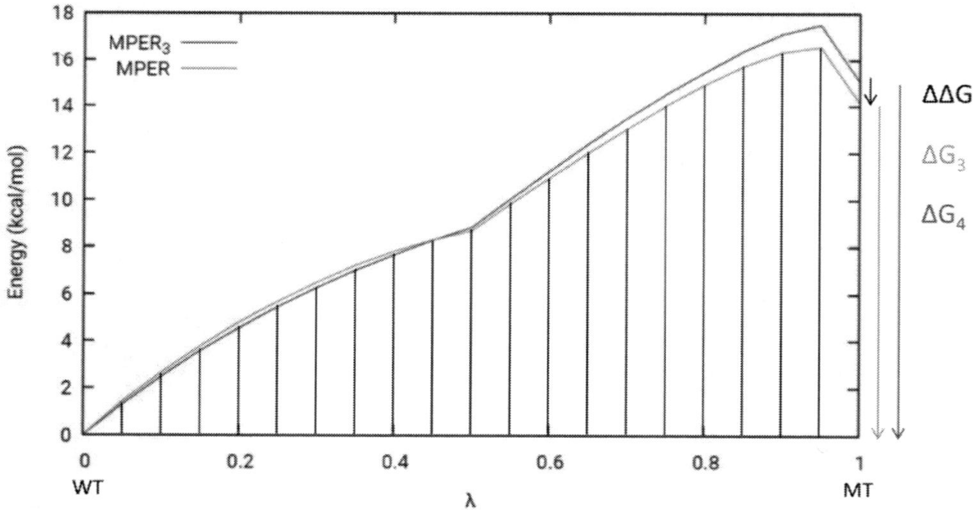

SOS equation

Lu, N.; Kofke, D.; Woolf, T. Journal of Computational Chemistry 2004, 25, 28–39.

$$\exp(-\beta \Delta A_{i\to i+1}) = \frac{\left\langle \exp\left\{-\frac{\beta}{2}\left[\mathcal{H}(x,p_x;\lambda_{i+1}) - \mathcal{H}(x,p_x;\lambda_i)\right]\right\}\right\rangle_i}{\left\langle \exp\left\{-\frac{\beta}{2}\left[\mathcal{H}(x,p_x;\lambda_i) - \mathcal{H}(x,p_x;\lambda_{i+1})\right]\right\}\right\rangle_{i+1}}$$

12

FEP shows that chain C is unique in the trimer structure in that only W666 is essential while W672 is essential on partner chains.

Mutation	Chain A ΔΔG (kcal/mol)	Chain B ΔΔG (kcal/mol)	Chain C ΔΔG (kcal/mol)
W666A	1.62±0.72	1.63±1.13	1.50±0.77
W670A	0.87±0.75	0.97±0.63	0.70±0.55
W672A	1.30±0.73	1.46±0.35	0.21±0.29
W678A	0.27±0.17	0.07±0.14	0.15±0.14

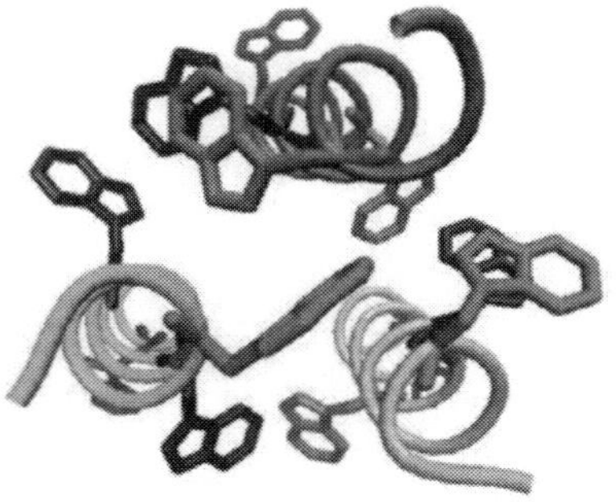

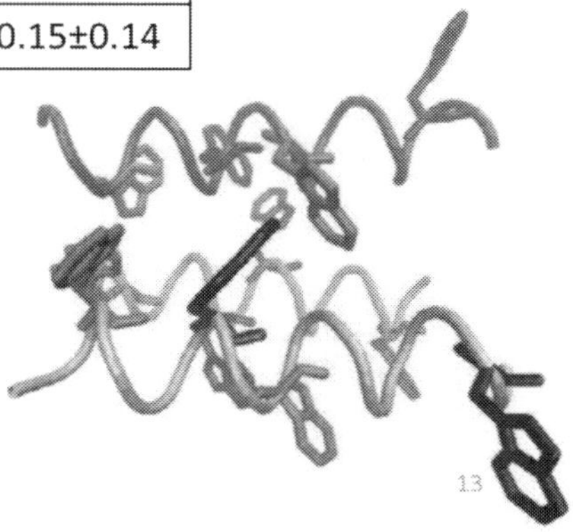

13

SASA shows a similar trend on chain C both for averages and in crystal structure as the FEP results.

Residue	Chain A SASA (Å^2)	Chain B SASA (Å^2)	Chain C SASA (Å^2)
W666	83 ± 5 (59)	103 ± 16 (105)	79 ± 35 (21)
W670	99 ± 8 (84)	102 ± 16 (107)	95 ± 2 (112)
W672	87 ± 16 (64)	99 ± 6 (106)	117 ± 17 (131)
W678	161 ± 2 (146)	149 ± 11 (168)	151 ± 3 (139)

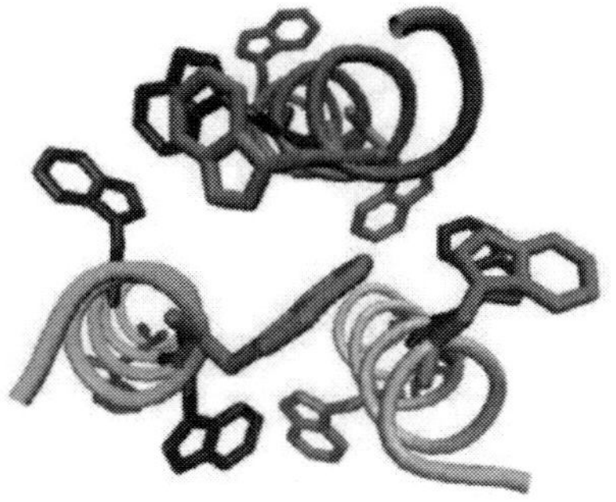

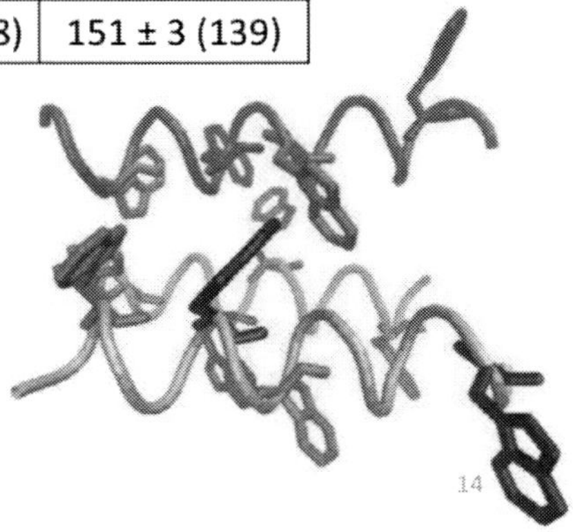

14

SASA for each individual simulation shows minor correlation between ΔΔG and SASA.

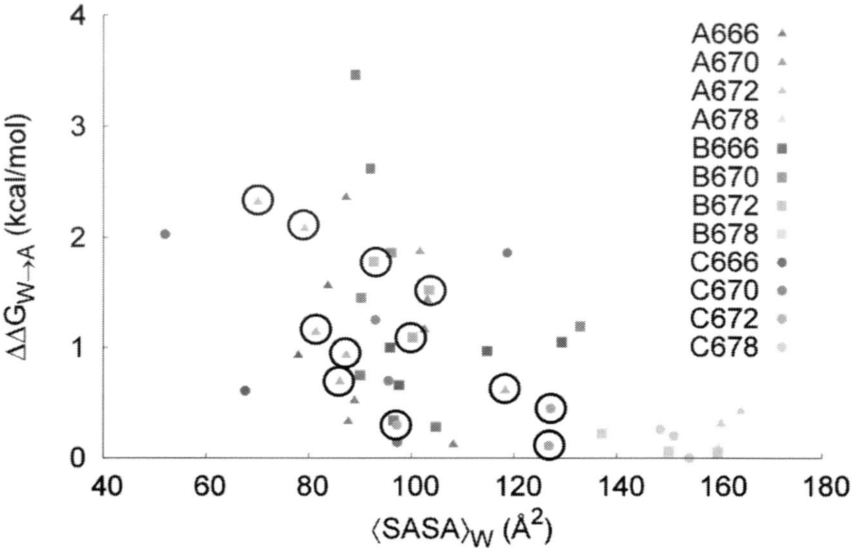

15

Symmetry of trimer can be tested by rotating chain A onto other chains.

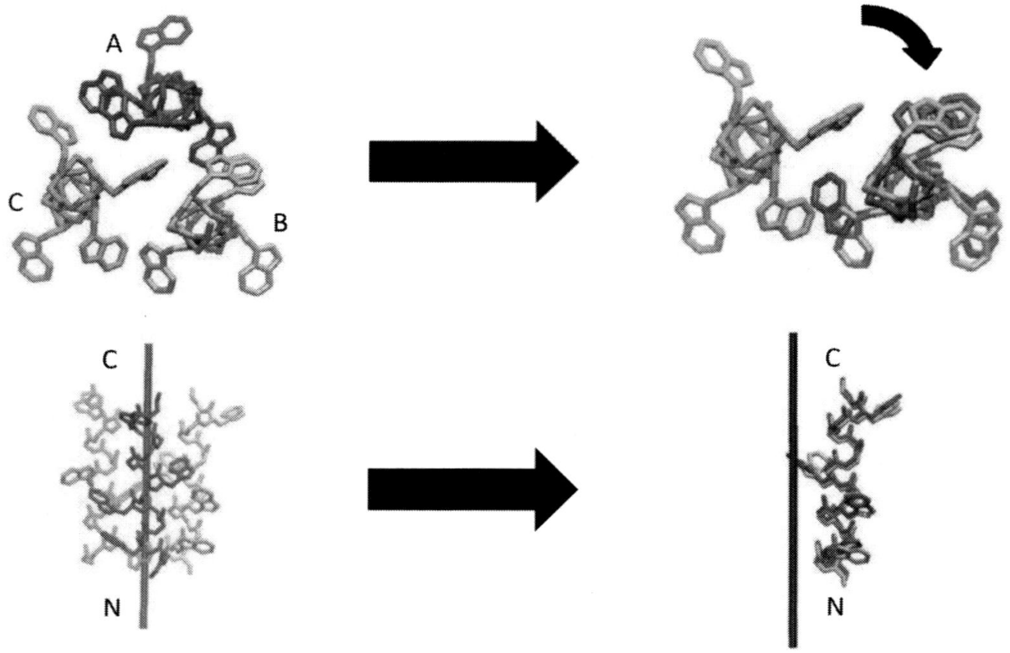

16

Symmetry of trimer can be tested by rotating chain A onto other chains.

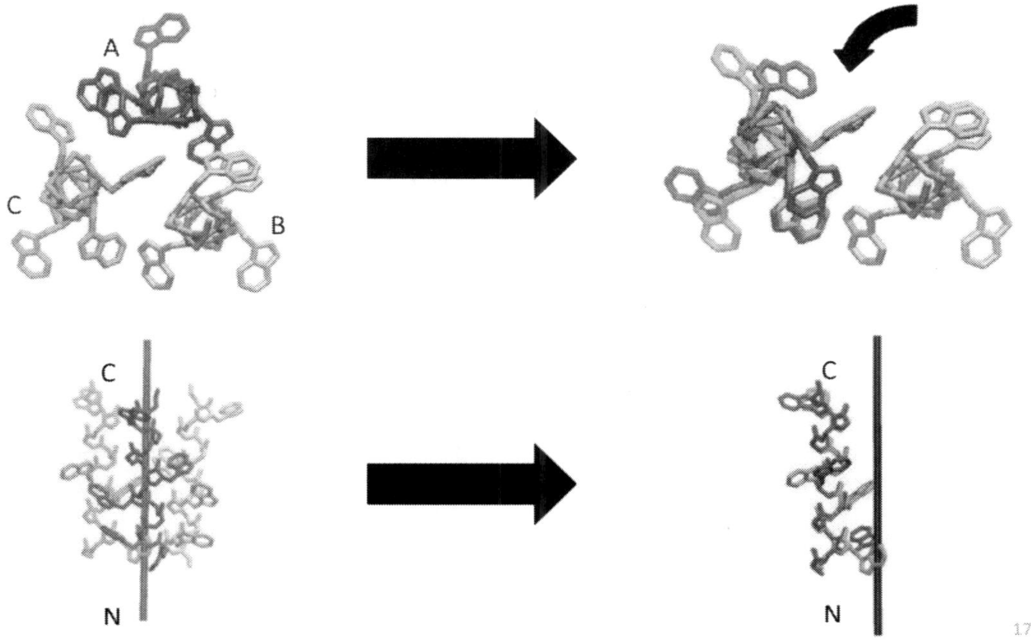

17

Due to chain C W666 asymmetry W672 on chain C is further away from trimer than complementary W672.

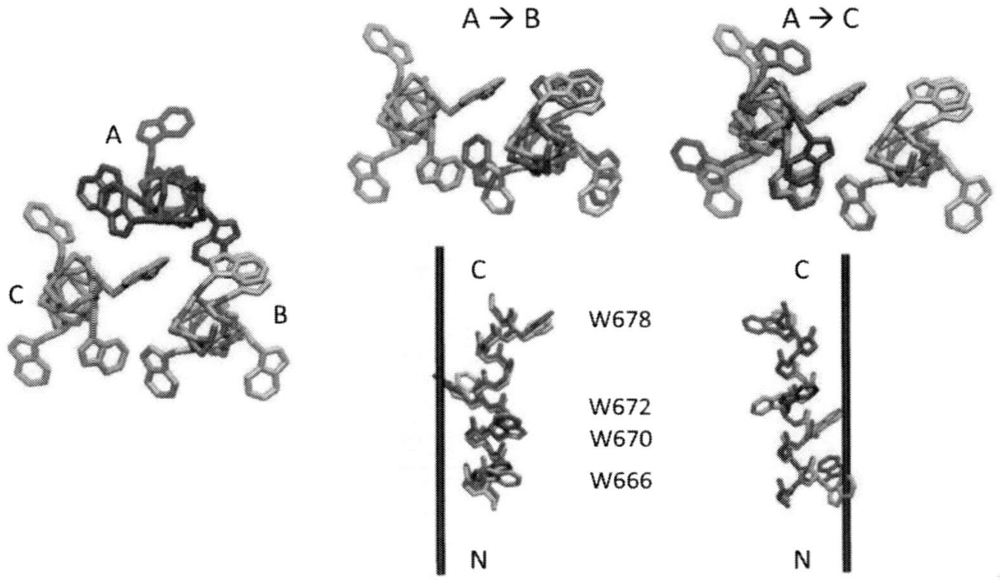

18

Conclusions

- W666 is essential on all chains while W672 is only essential on chains A and B
- Difference in chains is due to asymmetric trimer structure
- This model suggests that chain C may represent DAVEI-MPER interaction with native Env-MPER

19

Future Work

- Mutations of other residues in DAVEI sequence
- Mutations to Env-MPER
- Simplifying DAVEI molecule
- Funding provided through:
  - NIH R01 GM115249

20

Dual Topology Approach

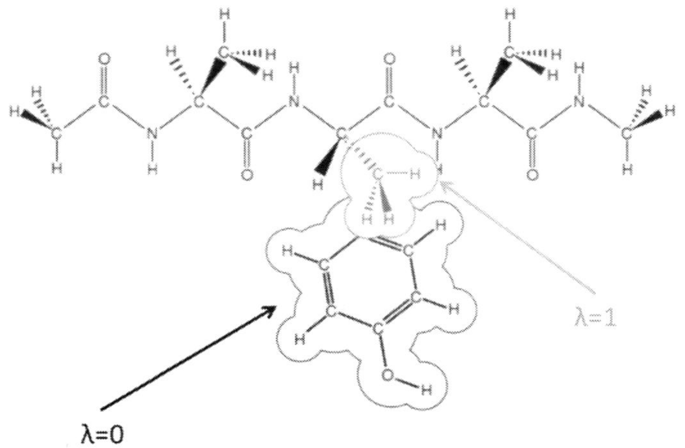

J. Henin, J. Gumbart, C. Chipot. In silico alchemy: A tutorial for alchemical free-energy perturbation calculations with NAMD. Universite de Lorraine. 2014.

21

FEP generates a ΔG for a mutation across λ using a forward and backward mutation.

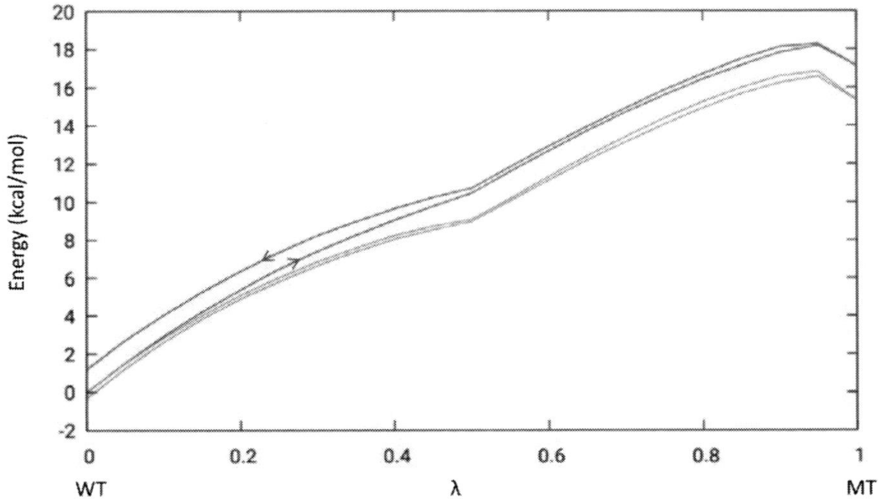

$$\Delta A_{\lambda \to \lambda + \partial\lambda} = -k_B T \ ln \left\langle exp \left[-\frac{\mathcal{H}_{\lambda+\partial\lambda}(r,p) - \mathcal{H}_{\lambda}(r,p)}{k_B T} \right] \right\rangle_{\lambda}$$

22

Effects of Asymmetry-Solvent Accessible Surface Area (SASA)

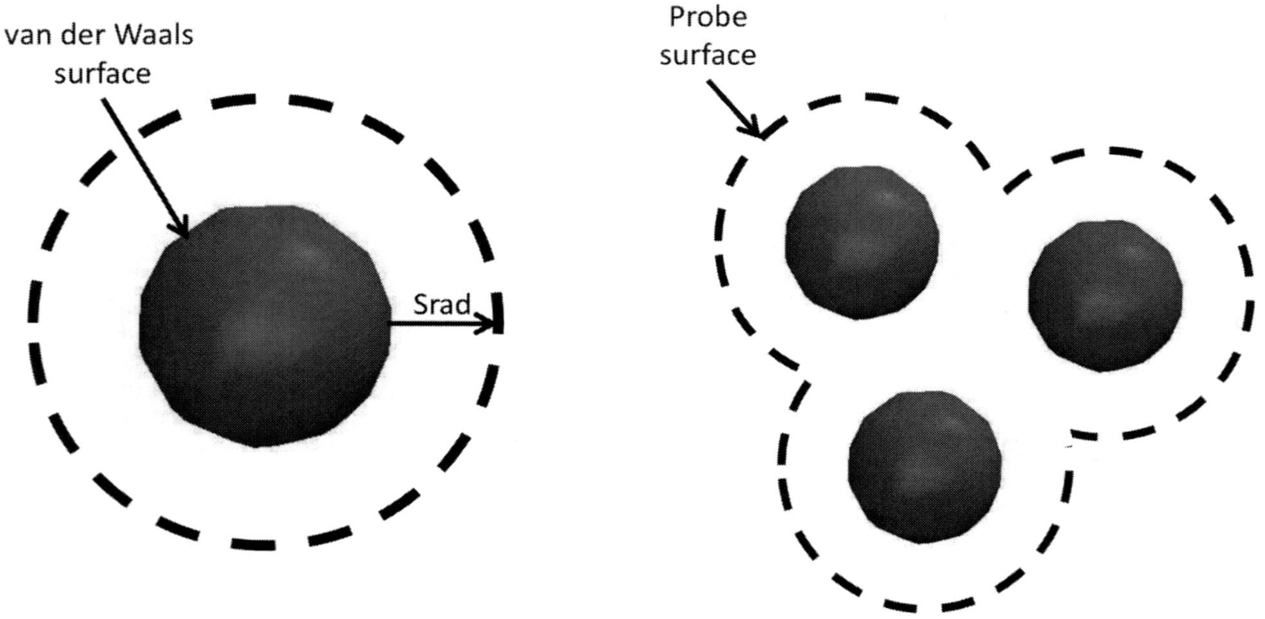

23

496920 Mathematical Modeling of Ultrasound in Regenerative Medicine:Â from the Cellular Scale to the Macroscale

Abstract:

As articular cartilage is avascular with limited ability for self-repair, and as a result osteoarthritis and other cartilage injuries are biomedical burdens (Cole, Pascual-Garrido et al. 2010, Huey, Hu et al. 2012). Current clinical and experimental treatment methods that aim to restore articular cartilage and heal full-thickness chondral defects are often characterized by functionally inferior fibrocartilage, hypo-cellularity and poor integration the repair site (Redler, Caldwell et al. 2012, Camp, Stuart et al. 2014). Ultrasound has become an indispensable tool in diagnostic imaging, is an FDA approved noninvasive therapy mechanism and is used as an operative tool. While high intensity US (1-3 W/cm2) finds utility in ablative applications, there is a paucity of published work that investigates the ability of low intensity US (LIUS, 1-50 mW/cm2) to improve cartilage repair outcomes and the efficacy of LIUS to repair articular cartilage injury, in vivo, remains unclear with conflicting outcomes (Cook, Salkeld et al. 2008, Yang, Kuo et al. 2014, Miller 2017). Thus we investigated low-intensity continuous ultrasound (LICUS) in cartilage restoration (Louw, Budhiraja et al. 2013). In a departure from all previous studies that employ low-intensity-pulsed-ultrasound (LIPUS) at 1.5MHz and 1kHz repeat; our work has established that the cell resonant frequency lies in the vicinity of 5 MHz for suspended and anchored cells (Louw, Budhiraja et al. 2013, Miller 2017). Further stimulation ofchondrocytes and mesenchymal stem cells seeded in hydrogels and macroporous scaffolds, in vitro, at the resonant frequency resulted in enhanced chondrogenesis, improved biochemical measures, biomechanical properties and histology scores when compared to control, non-stimulated cultures(Guha Thakurta, Kraft et al. 2014, Guha Thakurta, Budhiraja et al. 2015, Guha Thakurta 2016). Successful in-vivo translation of the positive bioeffects observed in vitro, necessitates overcoming two key challenges: (A) Identification of an in-vivo resonant frequency and (B) Propagation of LIUS in the joint space. Thus in order to develop a patient specific optimal therapeutic ultrasound treatment regimen that is rigorous, the following research objectives were undertaken: 1) the in vivo chondrocyte resonant frequency was identified; and 2) finally the extent of ultrasound propagation through the joint space was theoretically modeled and verified experimentally.

To elucidate underlying mechanisms in ultrasound induced bioeffects, we first showed that low-intensity ultrasound (LICUS) applied at resonance (~ 5MHz) induces deformation equivalent to that applied at 1MHz and significantly higher intensities (170kPa) (Miller 2017). The stored mechanical energy as a result of this deformation was maximized at resonance and the energy density in the nucleus was twice as high as in the cytoplasm. A mechanochemical model that linked the mechanical stimulation of ultrasound and the increased mechanical energy density in the nucleus to the downstream targets of the ERK pathway showed that ultrasound stimulation induces frequency dependent gene expression as a result of altered rates of transcription factors binding to chromatin. The bifurcation behavior of the cell when it is excited near resonance showed multiplicity. At positive detuning the mechanical energy coupled to the cell is small, it is higher at resonance but significantly higher at sub-resonant frequencies in the multiplicity range. Thus, there exists an optimal range of frequencies for ultrasound treatment where mechanical energy coupling is maximized (Miller 2017). As the resonant frequency is highly dependent on the mass and stiffness, we identified the resonant frequency of an in vivo chondrocyte to be in the range of 3.5-4.1 (Fig. 1D).

Limited research has been done to determine the extent of US propagation through the joint space. Only reported study we could find is the one by White and coworkers where they experimentally investigated the propagation of a 1.5 MHz frequency pulsed US regimen through the joint space of a cadaver human knee (White, Evans et al. 2007, White, Evans et al. 2010) and did not explore the use of higher frequencies. However, their model did not include the presence of tissue, cartilage, and meniscus to allow a better match to their experimental work. Due to greater attenuation in tissues rich in protein and the use of water, their model would overestimate the field strength (Fig 2, bottom black line). Thus, there is a need to develop a clinically translatable US regimen that takes into account properties of the cell and joint tissue. We investigated the extent of ultrasound propagation through the rabbit knee joint space and the sheep elbow space and coupled these experimental results with computer simulations of the joint spaces. The joint space leads to high attenuation thus the aid of computational models can assist clinicians in determining the effectiveness of ultrasound on patient specific cartilage restoration and optimal transducer positioning.

491124 Kelvin-Helmholtz Instabilities during Bacterial Separation from Blood

Abstract:

Using centrifugation, bacteria can be separated from whole blood due to the differences in sedimentation velocity. These sedimentation velocities and differences in density cause the red blood cells (RBCs) to form a separate layer from the blood plasma, which still contains many platelets and bacteria. This creates a two-phase system, with differences in both viscosity and density between the layers. Upon deceleration different velocity profiles are produced for each phase, which can cause instabilities (or mixing) to occur in the system. Under high centrifugation forces, these two phases will remain stable and separated from each other. However, as the system slows down to stop, mixing has been observed under some conditions. The mixing in this system arises from Kelvin-Helmholtz instabilities. We have performed both a mathematical analysis of the instabilities in the two phase flow, as well as an experimental analysis to verify the theory. The mathematical analysis was performed through a nonlinear energy stability analysis because of the velocity dependence on time. This analysis provides the criteria for stability, not instability. Once the energy equations were obtained, numerical methods were used to determine the disk velocity when the system leaves the stable region. This does not guarantee that the system will mix but means that the system may mix if other forces are out of balance. The analysis predicts that the boundary of the stable deceleration regime has a parabolic shape when plotting the rate of deceleration against the current velocity. The experimental analysis of stability produced the criteria for instability. This analysis was performed by slowing the blood down at constant rates of deceleration. This allowed us to determine the disk speed at which the blood would mix for different rates of deceleration. Experiments were performed with deceleration rates from 100 RPM/s down to 1 RPM/s. Because this analysis gives criteria for instability and the mathematical analysis gave criteria for stability, there is a gap between the two. In this gap, the interface between the 2 phases may go unstable and is dependent on the viscosity, density, and velocity differences as well as the centrifugation forces. Also, because RBCs are cells and not a fluid, the experimental analysis seems to deviate even farther from mathematical analysis which assumed the RBCs to be a fluid. Due to these differences, the experimental analysis predicts more of a straight line than a parabolic shape when plotting the rate of deceleration against the current velocity. Our hypothesis is that the mathematical analysis overcompensates for stability in the system in the mid-range deceleration rates (10 RPM/s â€“ 50 RPM/s).

Â For clinical applications, it may then be necessary to take the average between the mathematical and the experimental analyses so that the two-phase flow remains unmixed throughout the process and eliminate some of the overcompensation that the mathematical analysis predicts. This will lead to an optimized control of the system to allow for quick collection of bacteria from the blood.

491900 Dynamic Deformation of the Cell Plastically Shapes the Nucleus and Amplifies Cancer Nuclear Irregularities

Abstract:

Shapes of cell nuclei normally conform to the overall cell shape. This conformity is commonly interpreted as the stiff nucleus being elastically deformed in response to static cytoskeletal stresses whose distribution on the nuclear surface depends on cell shape. However, we found that deformed shapes of nuclei removed from cells by microdissection were preserved, both for smooth, elongated nuclei in fibroblasts and irregularly shaped nuclei of breast cancer cells. The lack of shape relaxation implies that nuclear deformations are plastic and the cellular stresses that deform the nucleus are dissipative, not static. Tracking changes of nuclear and cellular shape on micropatterned substrata revealed that fibroblast nuclei deform only during deformations in cell shape and only in the direction of nearby moving cell boundaries. Nuclear shape irregularities in MDA-MB-231 cancer cells increased during dynamic cell spreading, but decreased in MCF-10A cells. Furthermore, nuclear irregularities of cancer cells were reduced upon the disruption of the LINC complex. Elastic deformation of the cancer cell nucleus caused by a direct nuclear force probe did not correlate with its irregularity, nor with changes in irregularity upon perturbations to the nuclear envelope. We propose that the nucleus in a cell undergoing changes in shape is plastically deformed by dissipative stresses transmitted through the cytoplasm from the moving cell boundary to the nuclear surface.

492034 Combined Treatment of Heat and Antibiotics to Mitigate Biofilms on Implanted Devices

Abstract:

Thousands of people in the United States each year require the removal of their implanted device due to bacterial biofilm infections. Biofilms are notoriously difficult to treat due to many reasons including changes in gene regulation, decreased diffusivity of antibiotics to the bacteria from the extracellular polymeric substance, and decreased metabolism from nutrient limitations. Infected implants typically require invasive surgery to remove the implant followed by a long-term antibiotic treatment before re-implantation can occur. The use of an *in situ* thermal mitigation process in combination with antibiotic treatment could mitigate the biofilm without invasive surgery. To better understand the effect of the combined treatment, *Pseudomonas aeruginosa* biofilms were subjected to heat shocks at from 37 to 80 °C for 1 to 30 minutes all while exposed to either ciprofloxacin, tobramycin, or erythromycin at concentrations between 0.125 and 128 µg/mL. The biofilms were exposed to antibiotics for a total of 24 hours and heat shocked 4 hours into the antibiotic exposure to demonstrate the effect of a brief heat shock integrated with sustained antibiotic treatment. Following the complete antibiotic exposure time the remaining viable colony forming units (CFU) were quantified by serial dilution and plating of the homogenized biofilm. High temperatures and exposure times were successful at completely eradicating the 10^7 CFU biofilms on their own, however, biofilms subjected to milder heat shocks largely recovered by the end of the treatment time. The combined treatment of the heat and antibiotics proved to be synergistic in nature for a select window of temperatures and exposure times for all three antibiotics even at concentrations with no observable effect on their own. At 60 °C for 5 minutes and 70 °C for 1 minute the heat and any of the three antibiotics had multiple orders of magnitude reduction in viable bacteria with the most notable being erythromycin reducing the viable population by six orders of magnitude when neither heat nor antibiotic had an effect on their own. Through inductive heating on an implant deviceâ€™s surface heat could be applied without invasive surgery. The synergic effect of heat and antibiotics could obviate thousands of surgeries and greatly improve patient quality of life.

493306 Rapid Separation of Bacteria from Whole Blood for Sepsis Diagnosis

Abstract:

Sepsis is a severe blood infection that is caused by bacteria in the blood stream. If the bacteria are antibiotic resistant, the mortality rate of this infection is very high. The current clinical methods to diagnose sepsis require culturing the blood sample prior to diagnostic procedures to identify the bacteria and their resistance profile. Culturing the blood samples is a time-consuming step and the mortality rate of the sepsis increases as the time passes. It is crucial to find methods of diagnosing sepsis that do not need culturing the samples. Identification of bacteria is possible using molecular diagnostics (DNA analysis), but since blood cells interfere with these types of analyses, they must be separated from the bacteria first.

We have developed a method for rapid separation of the bacteria from whole blood based on sedimentation velocities in a centrifugal field. Various blood cells and bacteria have different sizes and will have different sedimentation velocities. Spinning blood in our hollow disks creates 2 layered phases: a layer of blood cells and a layer of plasma. While blood cells and bacteria have similar densities, blood cells sediment 30-fold or more faster than bacteria. The red cells and white cells will quick form a layer at the outer wall of the spinning hollow disk and the slower bacteria will remain in the plasma. By carefully reducing the spinning speed, we are able to separate the plasma containing the bacteria from rest of the blood cells.

This presentation describes our research and results in studying the parameters of hollow disk size, spinning speed, and spinning time on removing cells and retaining bacteria in the plasma. Our results show that we are able to separate and recover about 50 percent of the bacteria from 7 ml of whole blood in about 1 minute spinning followed by careful deceleration of the disk. More than 95% of the cells are removed from the plasma. This spinning disk technology can be easily implemented in a clinical setting.

500254 Characterization of TiO_2 Nanotubular Sensor for Detecting Tuberculosis Volatile Organic Compounds

Abstract:

One of the deadliest disease in the world is Mycobacterium tuberculosis (MTB)[1]. In 2015 there were 10.4 million incident TB cases worldwide[1]. Lack of low cost point-of-care methods for screening/diagnosis of the disease is one of the major issues contributing to this world wide spread disease. MTB microorganism releases two unique volatile organic compounds (VOC) as byproducts which are methyl nicotinate and p-anisate. Having a low cost device that can detect the presence of these two unique VOCs rapidly and accurately can help minimize the number of patients affected.

To take advantage of the presence of these unique VOCs, a hierarchical cobalt functionalized Titanium oxide nanotubular substrate has been made. In our previous work, we showed that at certain biased voltages methyl nicotinate and p-anisate will bind with functionalized TiO2 nanotubes[2]. The sensor response is measured between the baseline current and the final current. Several variables affect the sensor response to VOCs, such as nanotubular morphology, the percentage of cobalt in the TiO2 nanotubes, the gas flow rate and both the concentration and volume of the VOC provided to the sensor.

In this work, we will present results of testing sensors with various morphologies at different volumes, flow rates and concentrations of the two biomarkers mentioned in order to characterize trends in sensor response. This will help us optimize our results from the field and measure the patientâ€™s stage of disease. Sensor morphology and the amount of cobalt on TiO2 nanotubes will also affect the baseline current of the sensor. In order to improve the sensorâ€™s performance, we will examine a range of baseline currents and its influence on reliability of the response and current signal to noise ratio. Detailed analysis of the sensor performance to different variable ranges will be presented.

1. Tuberculosis (TB) Data and Statistics. https://www.cdc.gov/tb/statistics/ Accessed 12/16/2016, 2016.
2. Bhattacharyya D, Smith YR, Misra M, Mohanty SK. Electrochemical detection of methyl nicotinate biomarker using functionalized anodized titania nanotube arrays. *Materials Research Express* 2015;**2**(2):025002.

501467 Novel Reverse Electrodialysis Biofuel Cell

Abstract:

Biofuel cells offer the potential for a continuous, implantable source of power. A primary area of focus for biofuel cells has been the use of glucose oxidase immobilized enzymes. While in theory this enzyme has the potential to produce power for long periods of time inside of the body, in practice the immobilized enzyme breaks down over the course of anywhere from hours to days. Therefore, there is a need to explore other technologies to realize the goal of a long term biofuel cell. The objective of this research was to use a completely new approach to biofuel cells by using Gibbs Free Energy of Mixing between a low concentration and a high concentration stream to generate power. Within the body, the blood in the renal vein is reduced in ions (low concentrations) due to the waste removal function of the kidneys. Thus mixing the renal vein's blood with blood in another vein can produce power by reverse electrodialysis (RED). The RED-based biofuel cell has a power density on the same order of magnitude as the best glucose oxidase biofuel cells. Furthermore, the RED-based biofuel cell was operated for several weeks with little loss in power. In addition, the reduction in power due to the presence of the bovine albumin serum (BSA) protein was also investigated. Overall, the RED-based biofuel cell offers a consistent and sustainable method for generating energy to power medical devices.

497674 Nanocomposite for Implantable Electronic Devices

Abstract:

Traditional implantable electronic devices often induce inflammation due to the mismatch of the mechanical properties between the tissue and devices. Mitigation of this problem requires a new generation of medical devices fabricated from flexible and conductive materials. However, materials with such mechanical and electrical properties are not yet available in the toolbox of medical device fabrication. In this study, we formalized the mechanical and electrical requirements for implantable electronics. Then we utilized a nanocomposite approach to design and optimize the materials properties with various nanomaterials. Lastly, we fabricated implantable electronic devices using MEMS-like technology and demonstrated the functionality of the devices. Further investigation of the inflammation processes around these devices are required to establish their prospects as long-term implantable devices.

492935 Wireless, Battery-Free Optofluidic DeviceÂ Â for Programmable Fluid Delivery and Optogenetics

Abstract:

Diseases of the central and peripheral nervous systems, such as depression, anxiety, addiction, pain, Alzheimer's, amyotrophic lateral sclerosis and many others, affect the quality of life of millions of people worldwide. Progress in understanding the function of neural networks and in developing associated insights into the underlying causes of these mental health and neurological diseases can be accelerated by advances in techniques and methodologies for neuroscience research. Emerging devices for manipulation of neural activity in the deep brain create important new avenues for basic research into the operation of neural circuits, means of intracellular signaling, mechanisms for gene expression and other relevant biological processes. Here, we will present a wireless, battery-free optofluidic device that combine soft microfluidic and μ-ILED probes for programmable fluid delivery and optogenetics. This study yields an affordable technology for broad distribution to the neuroscience community, with a revolutionary unified set of capabilities in targeting specific neuronal populations in freely-moving animals over an unlimited period of time. In addition, the same platforms can easily be adapted for a wide range of other types of passive or active electronics functions, including electrical stimulation.

501944 Biomimicry in a High Cell Population Density Perfusion Centrifugal Bioreactor

Abstract:

There are many cellular and physiological processes that have been difficult to mimic within in vitro medical devices. We have developed a perfusion centrifugal bioreactor and demonstrated utility for a wide range of applications. In this talk we will give both a historical context as well as up-to-date new applications for use of such a device. Originally the centrifugal processor showed important use for separation of cellular components. Next it was applied to high population density cultures where it is useful in overcoming limitations of other processes by allowing the ability to provide fresh culture medium at high flow rates while retaining cells in suspension within a conical fluidized bed bioreactor stabilized by high gravitational forces. Applications included large-scale culture of hybridomas for production of monoclonal antibodies. The process is now being extended to bio-manufacturing of cytotoxic CD8 T-lymphocytes with applications in cancer therapy. Adding means for providing oscillating hydrostatic pressure, the system is showing use for the culture of anchorage dependent chondrocytes mixed with stem cells for production of more robust cartilage by stimulating chondrogenesis. Finally, the system can be used for elimination and/or detection of contaminating bacteria.

504788 Producing Protein Therapeutics without Cells

Abstract:

Recent engineering advances in in vitro or "cell-free" protein synthesis systems are enabling new possibilities for protein therapeutic production. Here we describe the development of lyophilized cell-free systems which provide the ability to economically stock-pile reagents for rapid response against an emerging threat and also enable facile transportation for just-add-water therapeutic production potentially anywhere. Importantly, the lyophilization processes also sterilizes the reagents which enables the use of genetically modified organisms to create cell-free reagents but prevents the release of genetically modified organisms to the environment during transit or storage. Additionally, a cell-free protein synthesis approach enables the production of complex, difficult-to-fold, and cytotoxic proteins at high production yields, solubilities, and activities. A case study producing Onconase (a promising cancer and anogenital wart biotherapeutic) using a lyophilized cell-free system is presented.

504247 *In situ* Thermal Eradication of Biofilms

Abstract:

Bacterial biofilms colonize 2 - 4% of the medical devices implanted in the U.S. each year despite rigorous efforts to maintain sterile operating conditions and extensive research into bacteria-resistant biomaterials. These biofilms are notoriously difficult to treat, requiring explantation of the implant and surrounding tissue, followed by eventual replacement of the device, at a cost of billions of dollars per year. Current strategies to develop biomaterials which prevent bacterial adhesion and colonization have inherent limitations which have prevented translation into devices with lower infection rates. We are pursuing a new approach to dealing with these infections, thermally eradicating them in situ using an alternating magnetic field. This requires multiple intersecting lines of investigation, from materials development to cell death kinetics to thermal modeling, each relying substantially on chemical engineering principles. This talk will discuss the thermal dose required for biofilm eradication, the energy requirements to supply that dose and the physiological consequences of that energy, and the design of coatings which are capable of delivering that dose from a given magnetic field source, as well as development of new tools to tackle these problems, such as pourable, size-stable thermal tissue phantoms.

A Single-use Sensor Array for Monitoring Growth Medium Analytes During Biomanufacturing of Monoclonal Antibodies

J.J. Magda, PhD
Professor, The University of Utah

THE UNIVERSITY OF UTAH
APPLIED Biosensors
NSF

Disclaimer: Professor Magda has financial interest in Applied Biosensors (AB) and is an equity stockholder at AB.

Outline of Presentation

- Current State of Biopharma Manufacturing
  - Shift toward single-use modular equipment
  - Batch mentality; lack of in-line sensors
- FDA Initiatives Designed to Modernize Biopharma Manufacturing
- Industry Requirements for New In-Line Sensors Needed for Bioreactors
- Applied Biosensors' Single-Use Sensor Array
  - Magnetic Smart Hydrogels
  - Contact-Free Signal Transduction
- Preliminary Results
- Conclusions

Shift in Biopharma Industry- From Stainless Steel Bioreactors Toward Single-Use-Bioreactors (SUBs) (reference C& EN News, Nov 23, 2015)

- Traditional Stainless Steel Bioreactor

VS.

- Modern Style Single-Use Bioreactor

The WAVE Bioreactor 20/50 system with WAVEPOD II integrated controller

<u>Advantages of Single-Use Bioreactors:</u>

1. Smaller initial capital investment
2. Flexibility with respect to production scale
3. Reduced water consumption – do not need to sterilize reactor after each production run
4. Decreased risk of cross-contamination between successive production runs

NOTE: Single-use bioreactors need single-use sensors

6

Biopharmaceutical Industry: *Elegant* Product Development, *Wasteful* Manufacturing

"Right now, manufacturing experts from the 1950s would easily recognize the pharmaceutical manufacturing processes of today" — *Janet Woodcock, Director of the FDA's Center for Drug Evaluation and Research, 2012.*

1950's bioreactor

Consider manufacturing of monoclonal antibodies (mAbs) – ($75 billion USD in 2013[#]):

- Performed using cells in a <u>batch</u> bioreactor
- The batch bioreactor has no in-line sensor for the principal reactant (glucose)
- The batch bioreactor has no in-line sensor for the principal metabolite (lactate)
- The batch bioreactor has no in-line sensor for the product (mAbs)

NOT SURPISINGLY:

- The pharmaceutical industry manufacturing at a lower efficiency (lower capacity utilization) than the petrochemical or food industrie

[#] D.M. Ecker, et al, mAbs (2015)
*R. Mullin, C&EN News (2007)

7

The FDA's *Process Analytical Technology Initiative* (PAT)

In 2004, the FDA released its PAT initiative in order to encourage voluntary adoption of Process Analytical Technology (PAT) in pharmaceutical manufacturing.

The Essence of PAT*

Process decisions are based on assessments of material attributes

- Forward-feed of incoming material
- In-process monitoring and control

Critical product attributes measured instantaneously (on-line, in-line, or at-line)

Current Technology: *Off-line* bioreactor sensors – sensor measurement time > time scale over which bioreactor properties change. Feedback control is impossible.

FDA-proposed technology: *In-line* bioreactor sensors – sensor measurement time < time scale over which bioreactor properties change. Feedback control is possible.

The Holy Grail – a continuous in-line sensor for measuring concentrations of mAbs in bioreactors

*C. Agarabi, www.fda.gov (2016)

APPLIED Biosensors

8

The BARDA-FDA *Continuous Manufacturing Innovations Initiative*

"Regulatory agencies, including the FDA .... support the adoption of *continuous manufacturing* for pharmaceutical production based on science- and risk-based approaches."*

Advantages of Continuous Manufacturing#

- Eliminates manual handling and human error
- Increases quality assurance *through online monitoring and control*
- Reduces manufacturing time and increases efficiency
- Reduces capital costs by using smaller equipment and less manufacturing space
- Responds more nimbly in the event of a drug shortage
- Allows manufacturers to tailor drug production to fit the needs of precision medicine

Continuous Biopharma Manufacturing ⟹ *Perfusion Bioreactor*

- Feedback loop to maintain steady state conditions
- It is impossible to maintain steady state without suitable in-line sensors

*S. Lee, FDA Center for Drug Evaluation & Research (2017)

 # Continuous Manufacturing: Improving Speed and Quality in Drug Production, FDA website (2017)

APPLIED Biosensors

9

Key Cell Growth Medium Properties for Which There are No Commercially Available In-Line Sensors

"Biopharma manufacturers are trying to implement better process monitoring, especially in single-use systems…..but the lack of reliable and effective sensors is inhibiting growth" (comment by E.S. Langer of Bioplan Associates in his 2015 industry survey)

For mAbs biomanufacturing, single-use sensors are needed for:

- Glucose
- Lactic acid
- pH
- Osmolarity

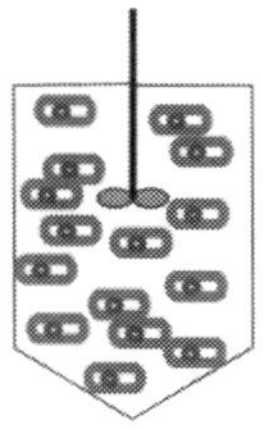

Fed-batch bioreactor containing genetically modified animal cells

Currently, there are no commercially-available sensors for these properties

Single-Use Sensor Requirements:

1. *Inexpensive* – single use
2. *Selectivity* – must function in chemically complex cell growth medium
3. *Large measurement range* – 1 to 50 mM for glucose, 6.5 to 7.5 for pH
4. *Robust* - can be sterilized via gamma irradiation & autoclaving
5. *Long shelf life* – must be able to sit on the shelf for six months prior to use

APPLIED Biosensors 10

Applied Biosensors' Continuous In-Line Bioreactor Sensor Array Based Upon Magnetic Smart Hydrogels

Smart Hydrogel = Crosslinked Polymer Network That Swells or Shrinks When It Reversibly Binds to Target Analyte

Example – *Glucose-Responsive Hydrogel*

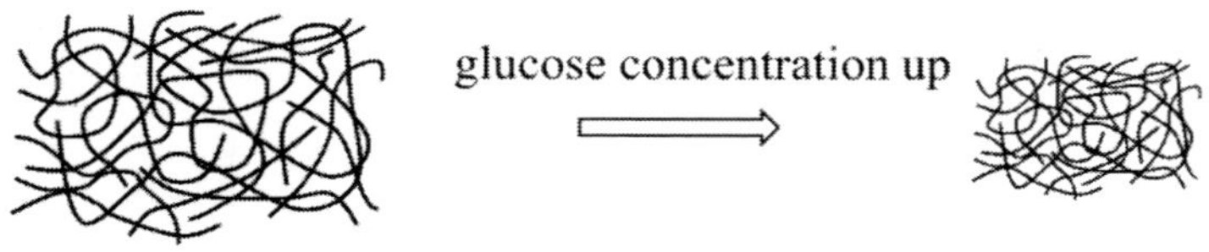

Four different smart hydrogels for 4 different target analytes: glucose, lactate, pH, osmolarilty

Hydrogel chemistry –polyampholytic acrylamide copolymer containing pendant phenyl boronic acid and tertiary amine moieties (Nguyen et al., *Sensors & Actuators B* (2018))

Sterilization stability – the hydrogels are stable to autoclaving and γ irradiation (Magda et al., *Polymer Degradation & Stability* (2014))

Continuous Sensor = Smart hydrogel + swelling transducer. We transduce swelling changes using a *magnetometer* to measure the change in magnetic field exerted by the hydrogel

APPLIED Biosensors 11

Contact-Free Signal Transduction Using *Magnetic* Smart Hydrogels

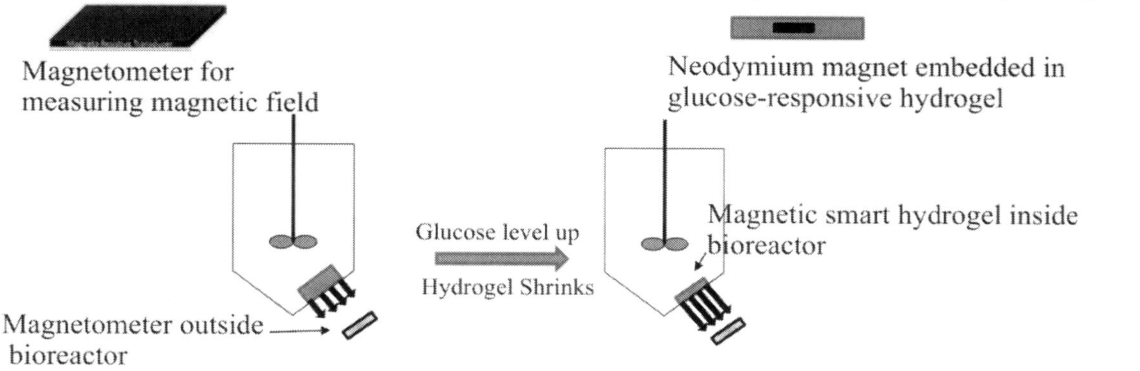

NOVELTY: Glucose monitoring is possible without inserting any electronics into cell growth medium

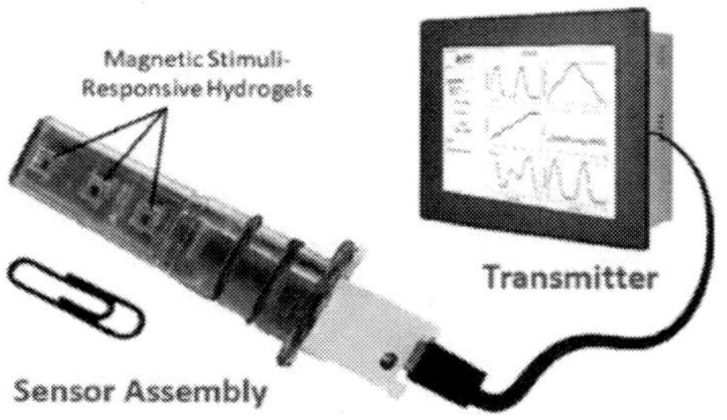

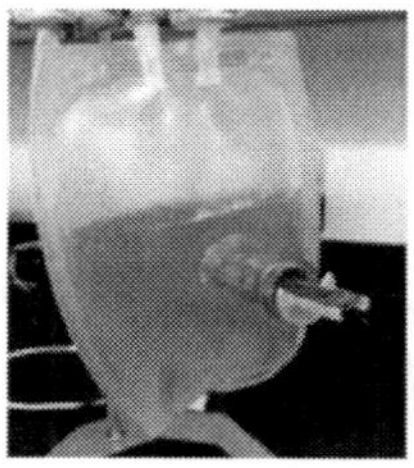

Sensor array inserts into bioreactor port

Response of Glucose Sensor to Cyclic Changes in Glucose in PBS

- As glucose concentration increases, the hydrogel shrinks, thereby increasing the magnetic field detected by the magnetometer

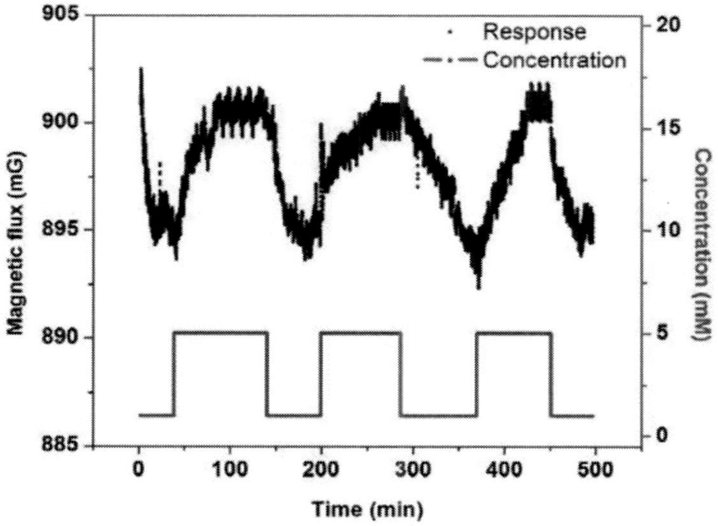

Glucose Sensor in a Commercial Cell Growth Medium

- The glucose sensor can be used to monitor glucose concentration in a complex cell growth medium over a very wide concentration range (1 – 50 mM)

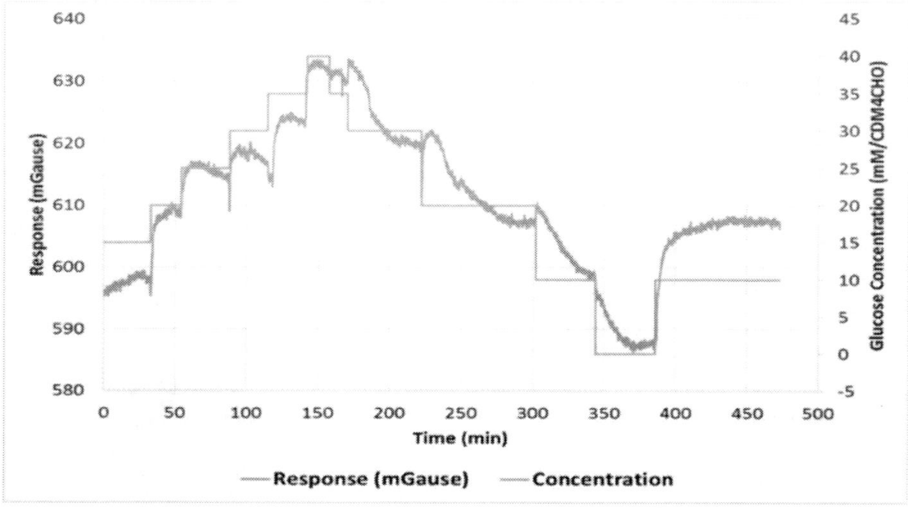

APPLIED Biosensors 15

Lactate Sensor in a Commercial Cell Growth Medium

- As lactate concentration increases, the hydrogel swells, thereby decreasing the magnetic field detected by the magnetometer

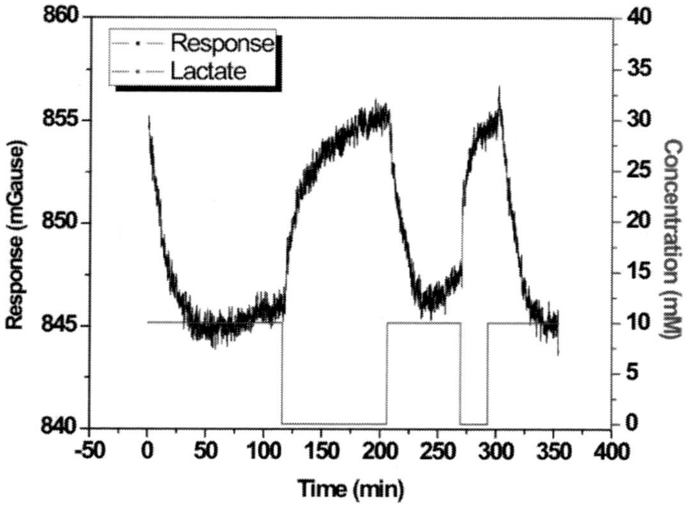

APPLIED Biosensors 17

Osmolarity Sensor in a Commercial Cell Growth Medium

- As osmolarity increases, the hydrogel shrinks, thereby increasing the magnetic field detected by the magnetometer

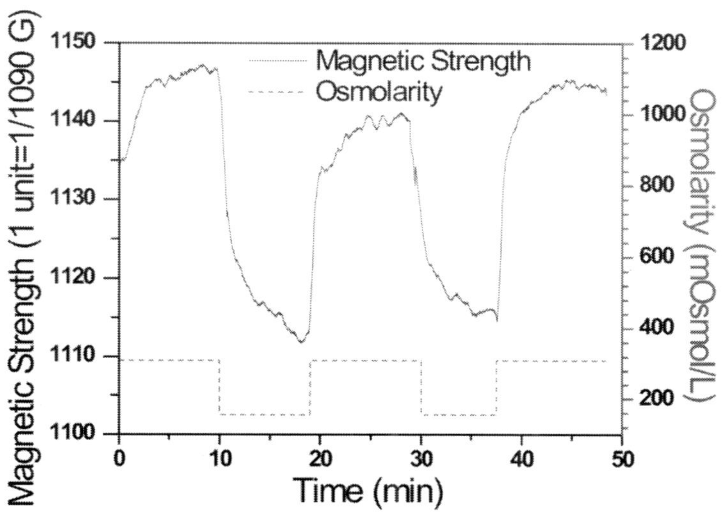

APPLIED Biosensors

Conclusions

1. Manufacturing in the biopharma industry is currently inefficient due to the lack of suitable in-line process sensors
2. The FDA is encouraging a shift in biomanufacturing towards
   - Use of single-use bioreactors
   - Use of more in-line process sensors
   - Use of continuous bioreactors
3. Suitable in-line sensors for bioreactors must be stable to gamma sterilization procedures. This rules out sensors that use antibodies or enzymes as recognition elements
4. However, sensors that use magnetic smart hydrogels as recognition elements are stable to gamma irradiation, and therefore have great potential for use as in-line process sensors during pharmaceutical manufacturing
5. Proof-of-concept has been demonstrated for bioreactor sensors based upon magnetic smart hydrogels

APPLIED Biosensors

Acknowledgements

- Applied Biosensors, LLC, Salt Lake City, UT
  - Dr. Seung-Hei Cho
  - Dr. Prashant Tathireddy
  - Mr. Rohit Sharma
  - Dr. Nick Frazier
  - Mr. Yoonsung Goo
  - Mr. Rich Barra and others

APPLIED Biosensors

- Financial Support (Partially)
  - National Science Foundation

National Science Foundation
WHERE DISCOVERIES BEGIN

498773 Numerical Study of Hemodynamics in the Carotid Artery before and after Angioplasty with Stenting Using Different Rheological Models

Abstract:

Abstract

The carotid arteries are the main source of the blood to the brain and any problem in its structure can cause harm to the individual. Among the disorders that may represent an irregularity there is stenosis, which is the narrowing of the blood vessel and one of the causes of stroke and transient ischemic attack. The angioplasty with stenting is one of the most used surgical procedures in the treatment of atherosclerosis, a disease that consists in the accumulation of fat in the walls of the blood vessel. Even presenting a severe carotid stenosis, the patient can be both symptomatic and asymptomatic, demonstrating the importance of monitoring the problem and its causes. In this context, computational fluid dynamics (CFD) has been employed as a powerful tool in the investigation of flow behavior in arteries and veins, especially in the characterization of areas predisposed to deviate from normality, such as stenosis. In the mathematical modeling, however, the representation of blood in arteries as a Newtonian or non-Newtonian fluid is still under study. In order to investigate the numerical difference arising from the modeling of blood viscosity in the hemodynamics of the carotid artery before and after angioplasty with stenting, geometries based on clinical exams of a patient were used. Thus, besides characterizing the numerical differences resulting from the change in rheological model, it was possible to highlight areas prone to the occurrence of stenosis and restenosis in each case.

497529 Investigating the Neuroprotective Effects of 5-Hydroxyadamantane-2-One on Middle-Aged Male Rats in an Ischemic Stroke Model

Abstract:

Introduction: Diamondoids are ultrastable saturated ringed hydrocarbons exhibiting antiapoptotic properties in neurodegenerative disorders including stroke. 5-Hydroxyadamantane-2-one (5-HA-2-one) is one of the derivatives of the simplest diamondoid molecule- adamantane- that is shown to have neuroprotective effects in global transient and local permanent ischemic rat models. However, unfortunately, no studies have evaluated the effectiveness of 5-HA-2-one either in aged animals or at a time point > 30 min following stroke. Therefore, this pilot study examined the neuroprotective effects of 5-HA-2-one on middle-aged male rats 6 hours after MCAo-induced ischemic stroke and compared it to the neuroprotective effects of this drug on mature adult male rats 1 hour and 30 minutes after stroke.

Methods: Mature adult rats (5 months old) and middle-aged (11 months old) Sprague-Dawley male rats were injected with endothelin-1 (ET-1, 600 pmol) stereotaxically toward the middle cerebral artery to induce occlusion (MCAo). 5-HA-2-one (100 mg/kg, i.p.) was administered at 1 hour and 30 minutes post ET-1 and once daily for 4 days to mature adult rats and 6 hour post ET-1 injection and once daily for 4 days to middle-aged rats. Sensory motor behavioral tests were performed pre (-2d) and post (2d and 5d) MCAo. Rats were sacrificed 5 days after MCAo. Brains were sectioned coronally and immersed in 2% 2,3,5-triphenyltetrazolium chloride in DPBS for 30 min at 37°C and processed for imaging. Serum and tissue samples were also collected for biochemical analyses.

Results: This pilot study (n=4) on Mature adult and (n=5) on middle-aged male rats with 5-HA-2-one treatment described above displayed no significant differences in infarct volume (p= 0.813) or on performance of sensory-motor tasks (p=0.866) on middle-aged rats but it showed a significant improvement in sensory-motor behavioral test (p=0.03) on mature adult rats when compared to post-stroke saline treated group. This observation differs from published reports that 5-HA-2-one exhibits neuroprotective effects on mature adult male rats in the early stroke phase 30 min after stroke.

Conclusions: This is the first report of the use of a diamondoid derivative in clinically relevant stroke prone group, the middle aged males, with delayed (> 6 hours) treatment. Our findings suggest that older males may require different strategy (e.g. higher dose and/or earlier treatment) in 5-HA-2-one treatment to exhibit neuroprotective effects. These data underscore the importance of testing potential therapeutic drugs in appropriate preclinical models.

Supported by NIH/AG042189 to FS

502167 Blood Damage Predictions Using Computational Fluid Dynamics of Blood Flow through a Bi-Leaflet Prosthetic Heart Valve

Abstract:

Artificial heart valves can be invaluable to addressing heart defects and diseases. However, these prosthetic devices expose the blood to unnatural turbulent flow conditions. This non-uniform flow exposes blood to high stress, which can damage the blood cells. The purpose of this research is to analyze blood flow in artificial heart valves and develop a new approach to predict the damage (hemolysis) caused to red blood cells (RBCs) due to the intensity and extent of turbulence. Using ANSYS ICEM CFD, a virtual bi-leaflet prosthetic heart valve was constructed and meshed. To compare the amount of damage in a working and malfunctioning valve, a similar valve, but with one closed leaflet, was also created. The simulations and calculations at different Reynold numbers and stress conditions are then done in ANSYS Fluent and validated with experimental findings in the literature. Results from the CFD simulations provide the spatial distribution of Kolmogorov length scales that are used for eddy analysis [1,2] in the heart valves. The Kolmogorov length scale (KLS), or the dissipative length scale, is used to find the size of eddies in the fluid. This CFD research enables approximation of the number and shape of eddies in the blood as a possible causation or correlation to the amount of damage to the blood cells.

Keywords: Computational Fluid Dynamics, Heart Valve, Blood Flow, Hemolysis, Eddy Analysis

References

1. M. Ozturk *et al.*, "Hemolysis Related to Turbulent Eddy Size Distributions Using Comparisons of Experiments to Computations",*Artificial Organs* (2015), 39(12): E227-E239.
2. M. Ozturk *et al.*, "An Approach for Assessing Turbulent Flow Damage to Blood in Medical Devices", *J Biomech Eng* (2017), 139(1), 011008.

500105 Intracellular Mass Transport Estimation Using Quantitative Phase Microscopy

Abstract:

Metastasis is the cause of 90% of cancer deaths. Cancer cells metastasis is a result of dysregulation of both mechanical properties and growth. Therefore there is a need for approaches which are capable of quantifying both cell mechanical properties as well as their growth in order to study cancer metastasis and to develop and deploy effective therapies. In this work we apply concepts and approaches from fluid mechanics to study cell mechanics based on quantitative phase imaging of living cells, a platform that has demonstrated utility in measurement of cancer cell growth.[1]

Quantitative phase microscopy (QPM) is a label-free, non-invasive imaging method which quantifies the total mass and mass distribution of single living cells. The principle of QPM is as follows: the combination of cell refractive index and thickness slows down light passing through a cell, resulting in a shift in phase relative to light passing through the surrounding media. This phase shift measurement can be used to compute cell mass using the known specific refractive index increment for cell biomass.[1,2] In biomedical applications, QPM data can be used to track total cell biomass over time to measure cell growth[3] or the decrease in growth rate caused by treatment of a cancer cell with an effective therapeutic.[4,5] QPM data also reveal the dynamics of cell mass motion, which is linked to cell mechanical properties.[6]

The focus of this work is the estimation of mass transport rate using QPM data. The cell is considered to be a viscoelastic fluid which obeys non-Newtonian behavior. The velocity of mass transport can then be estimated using particle tracking approaches common in experimental fluid mechanics. In this talk we will present our approach as well as the utility of this method to monitor mass redistribution and generation inside cancer cells as a measure of cancer cell growth and disease progression.

References

1. Zangle TA, Teitell MA. Live-cell mass profiling: an emerging approach in quantitative biophysics. *Nat. Methods.* 2014;11(12):1221-1228.

2. Barer R. Interference microscopy and mass determination. *Nature.* 1952;169:366-367.

3. Mir M, Wang Z, Shen Z, et al. Optical measurement of cycle-dependent cell growth. *Proc. Natl. Acad. Sci. USA.* July 25, 2011 2011;108(32):13124-13129.

4. Reed J, Chun J, Zangle TA, et al. Rapid, massively parallel single-cell drug response measurements via live cell interferometry. *Biophys. J.* 2011;101(5):1025-1031.

5. Chun J, Zangle TA, Kolarova T, Finn RS, Teitell MA, Reed J. Rapidly quantifying drug sensitivity of dispersed and clumped breast cancer cells by mass profiling. *Analyst.* 2012;137:5495-5498.

6. Eldridge WJ, Steelman ZA, Loomis B, Wax A. Optical Phase Measurements of Disorder Strength Link Microstructure to Cell Stiffness. *Biophys. J.* Feb 28 2017;112(4):692-702.

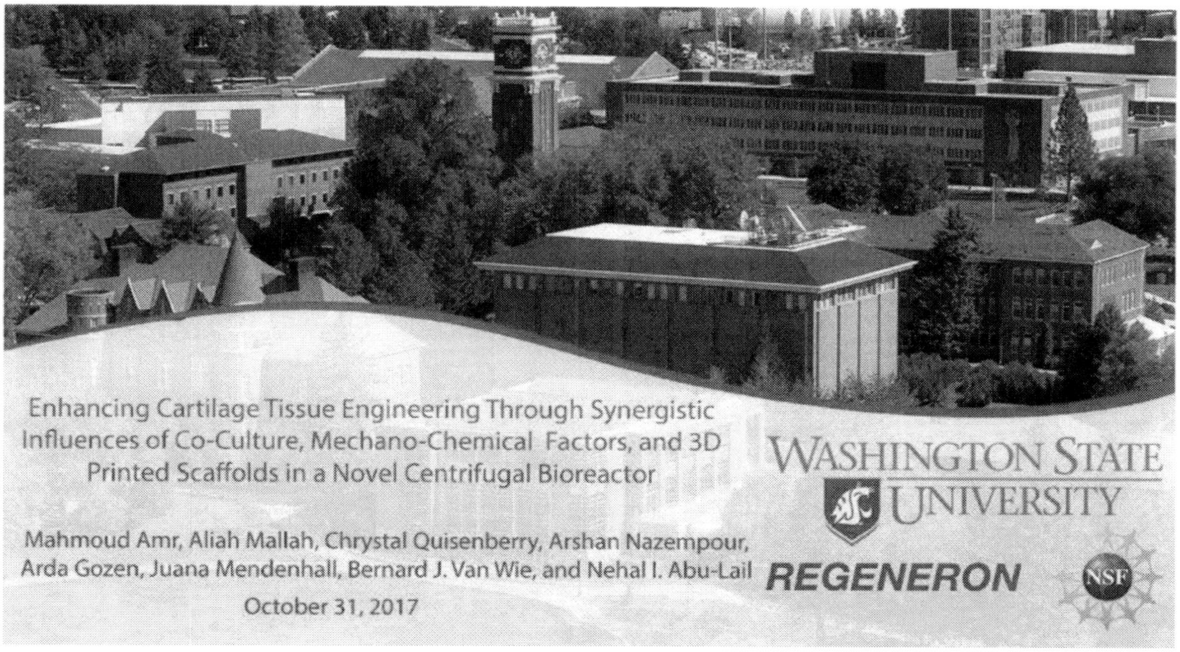

OUTLINE

- Overview
- Methods
- Results
- Future Directions
- Acknowledgments

2

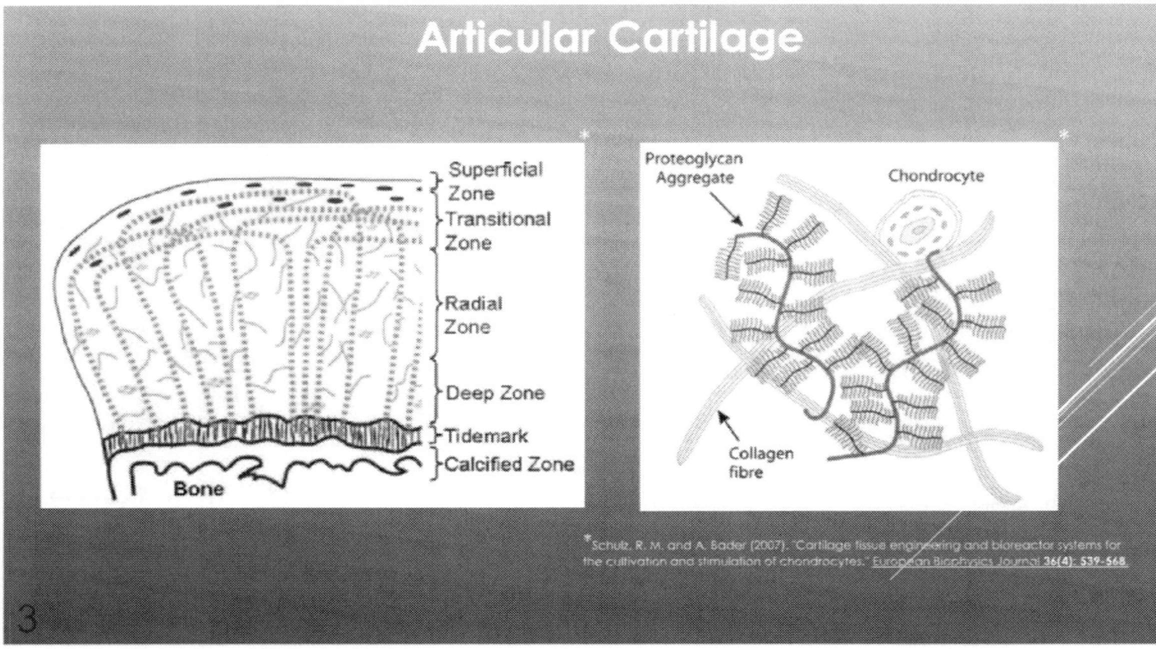

Articular Cartilage
Superficial Zone
Transitional Zone
Radial Zone
Deep Zone
Tidemark
Calcified Zone
Bone
Proteoglycan Aggregate
Chondrocyte
Collagen fibre
*Schulz, R. M. and A. Bader (2007). "Cartilage tissue engineering and bioreactor systems for the cultivation and stimulation of chondrocytes." European Biophysics Journal 36(4): 539-568.
3

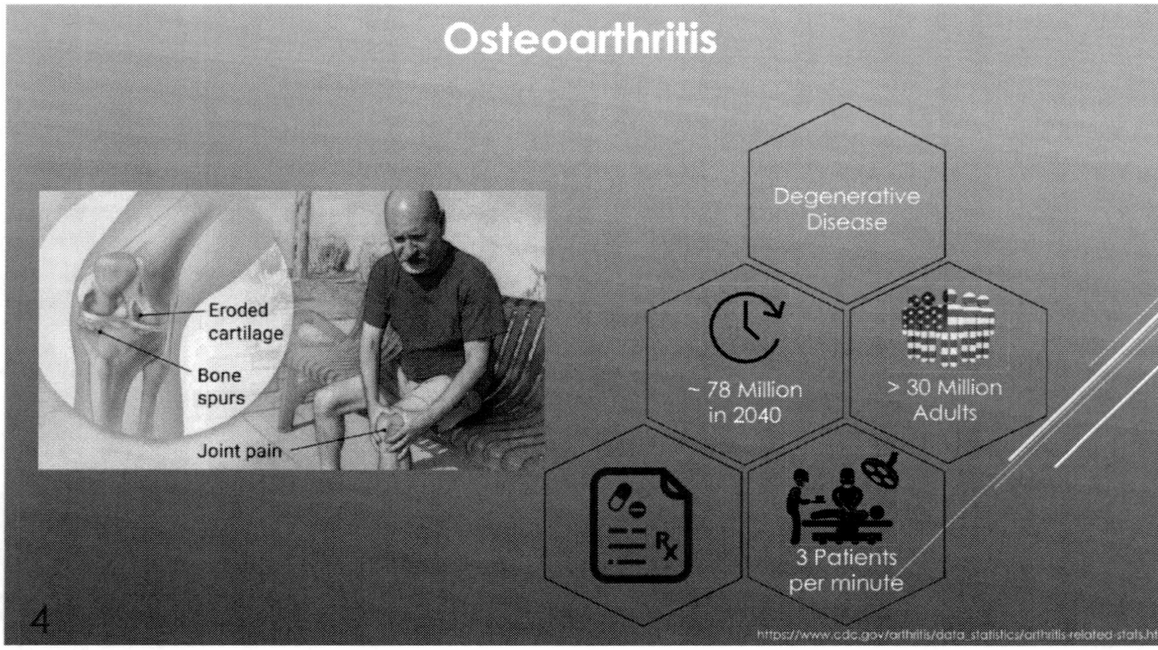

Osteoarthritis
Eroded cartilage
Bone spurs
Joint pain
Degenerative Disease
~ 78 Million in 2040
> 30 Million Adults
3 Patients per minute
https://www.cdc.gov/arthritis/data_statistics/arthritis-related-stats.htm
4

Articular Cartilage Tissue Engineering

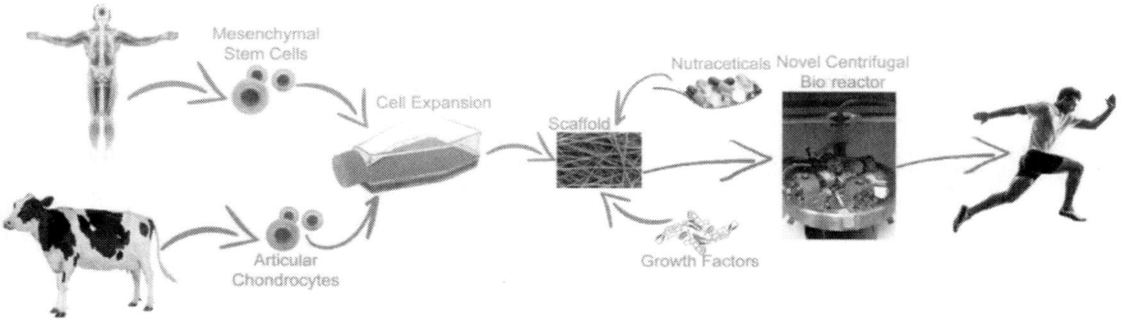

5

Our Goals

What is the effect of the following on chondrogenesis:

1. Oscillating hydrostatic pressure?
2. Growth factors?
3. Nutraceuticals?
4. Co-culture?
5. 3D printing?

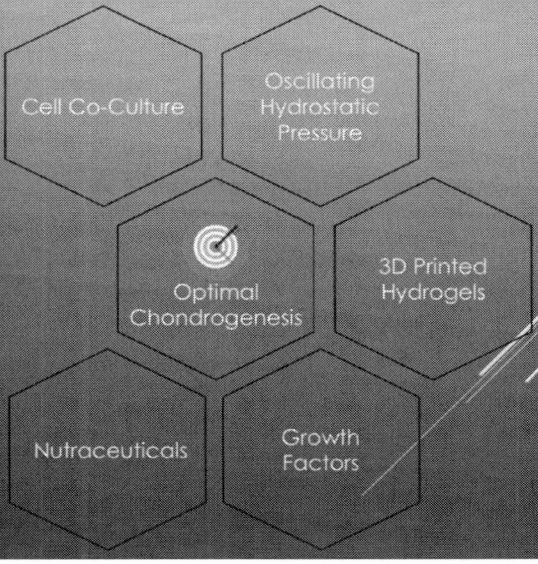

6

Centrifugal Bioreactor

- Force due to walking is 3.4 times body weight
- Oscillating hydrostatic pressure up to 5.5 MPa @ 1 Hz)
- Pulsatile perfusion bioreactor
- Centrifugal bioreactor
- Tissue is grown for 21 days

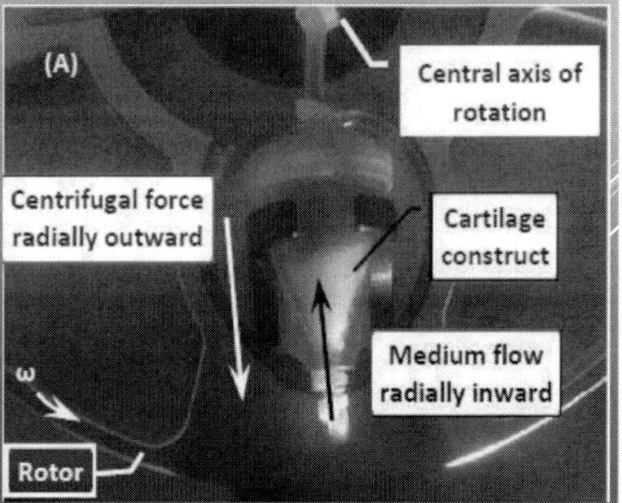

7

Characterization

- Gene expression using qRT-PCR
- Colorimetric test
- Immunohistochemistry and histology
- Biomechanics of cells and extracellular matrix using AFM

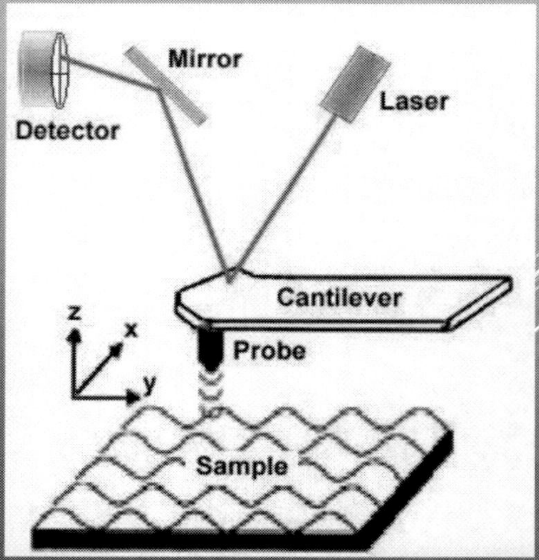

https://www.mtholyoke.edu/~menunez/ResearchPage/AFM.html

8

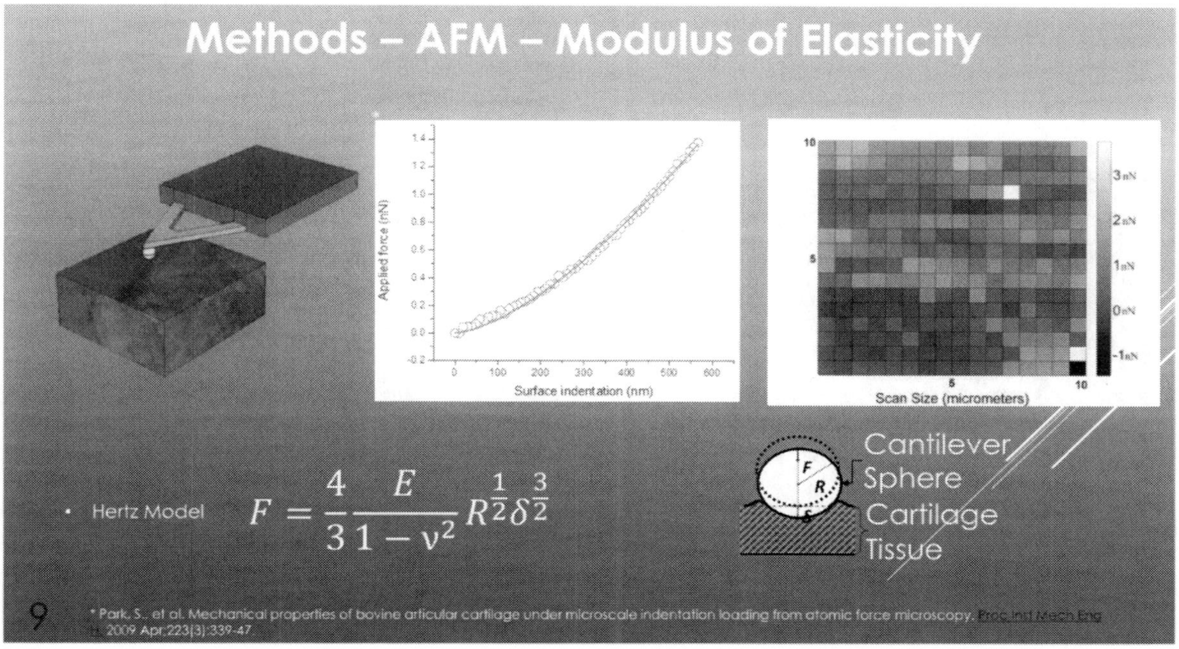
Methods – AFM – Modulus of Elasticity
Applied force (nN)
Surface indentation (nm)
Scan Size (micrometers)
3nN
2nN
1nN
0nN
-1nN
Cantilever
Sphere
Cartilage
Tissue
• Hertz Model $F = \frac{4}{3}\frac{E}{1-\nu^2}R^{\frac{1}{2}}\delta^{\frac{3}{2}}$
9
* Park, S., et al. Mechanical properties of bovine articular cartilage under microscale indentation loading from atomic force microscopy. Proc Inst Mech Eng H. 2009 Apr;223(3):339-47

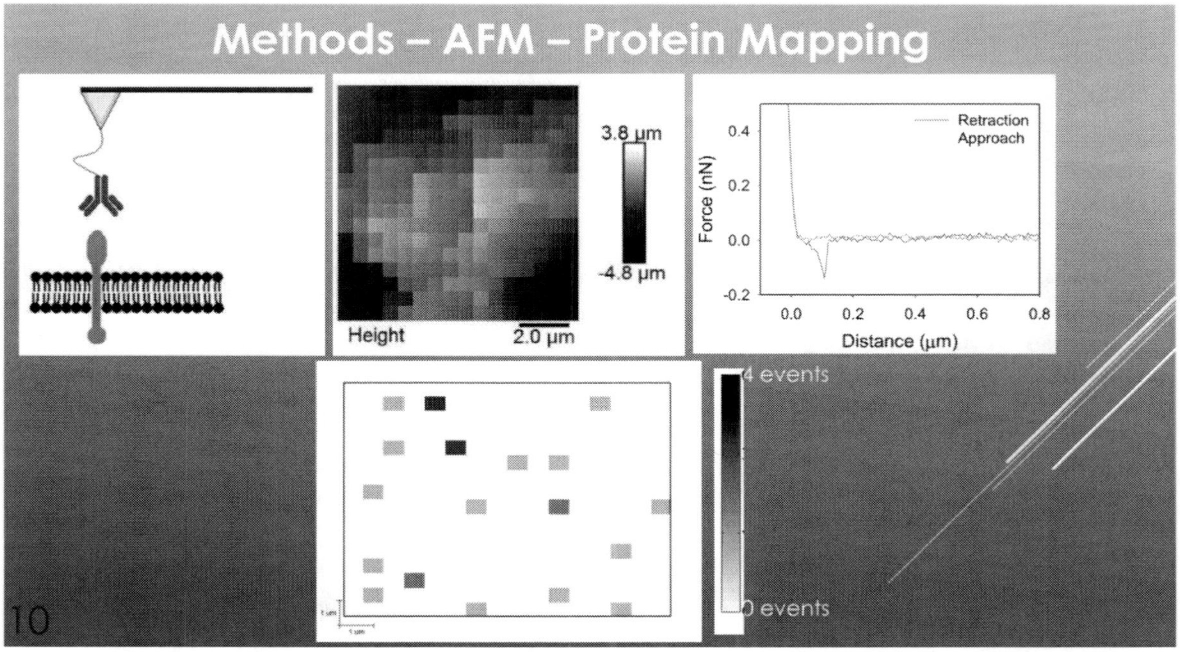
Methods – AFM – Protein Mapping
3.8 µm
-4.8 µm
Height
2.0 µm
Retraction
Approach
Force (nN)
Distance (µm)
4 events
0 events
10

Questions 1 and 2: what is the effect of oscillating hydrostatic pressure and growth factors on chondrogenesis?

11

Effects of TGF-β3 and oscillating hydrostatic pressure

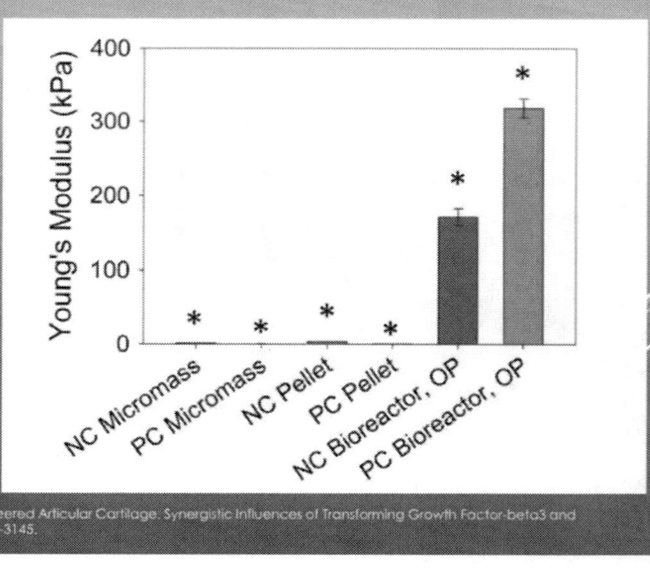

Human Articular Cartilage
Elastic Modulus: 450 – 800 kPa

12 Nazempour, A., et al. (2016). "Nanomechanics of Engineered Articular Cartilage. Synergistic Influences of Transforming Growth Factor-beta3 and Oscillating Pressure." J Nanosci Nanotechnol 16(3): 3136-3145.

Questions 1 and 2: what is the effect of nutraceuticals and growth factors on chondrogenesis?

13

TGF-β3 and Catechin - Total Collagen Histology

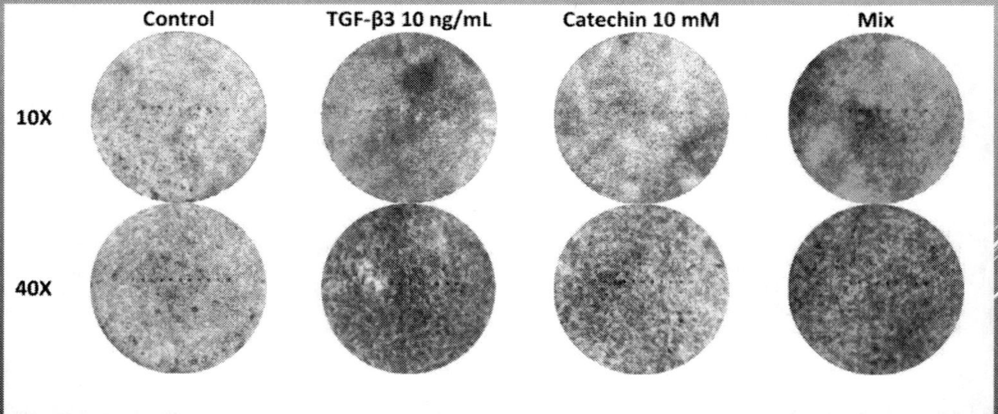

14

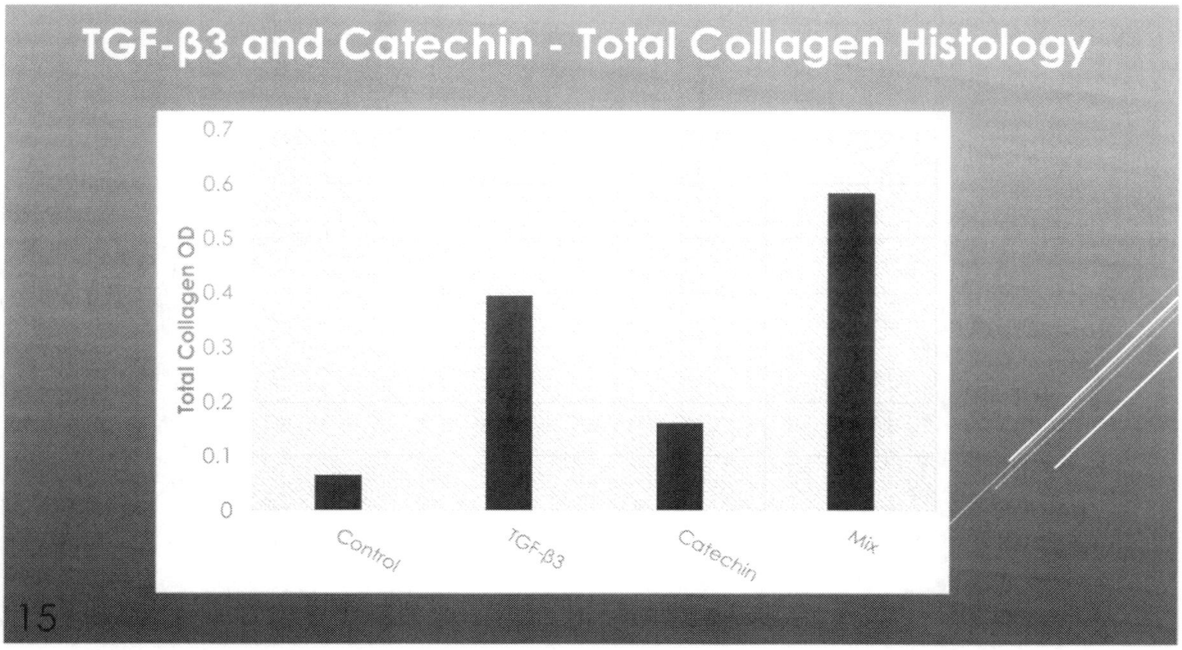
TGF-β3 and Catechin - Total Collagen Histology
0.7
0.6
0.5
0.4
0.3
0.2
0.1
0
Total Collagen OD
Control
TGF-β3
Catechin
Mix
15

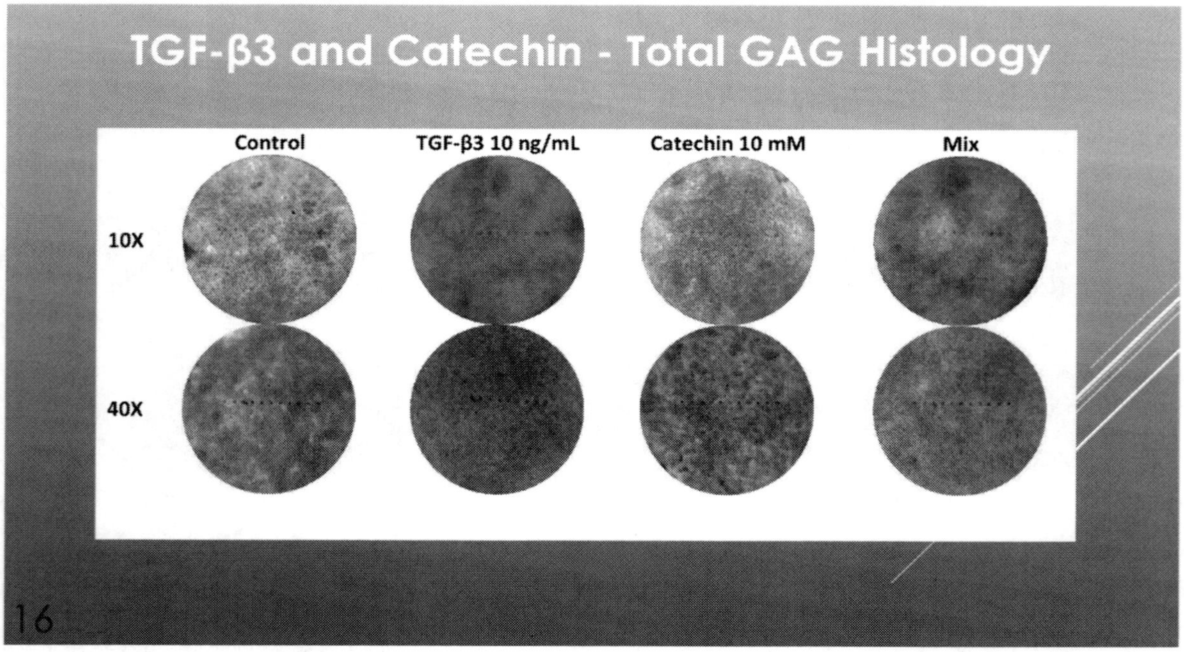
TGF-β3 and Catechin - Total GAG Histology
Control
TGF-β3 10 ng/mL
Catechin 10 mM
Mix
10X
40X
16

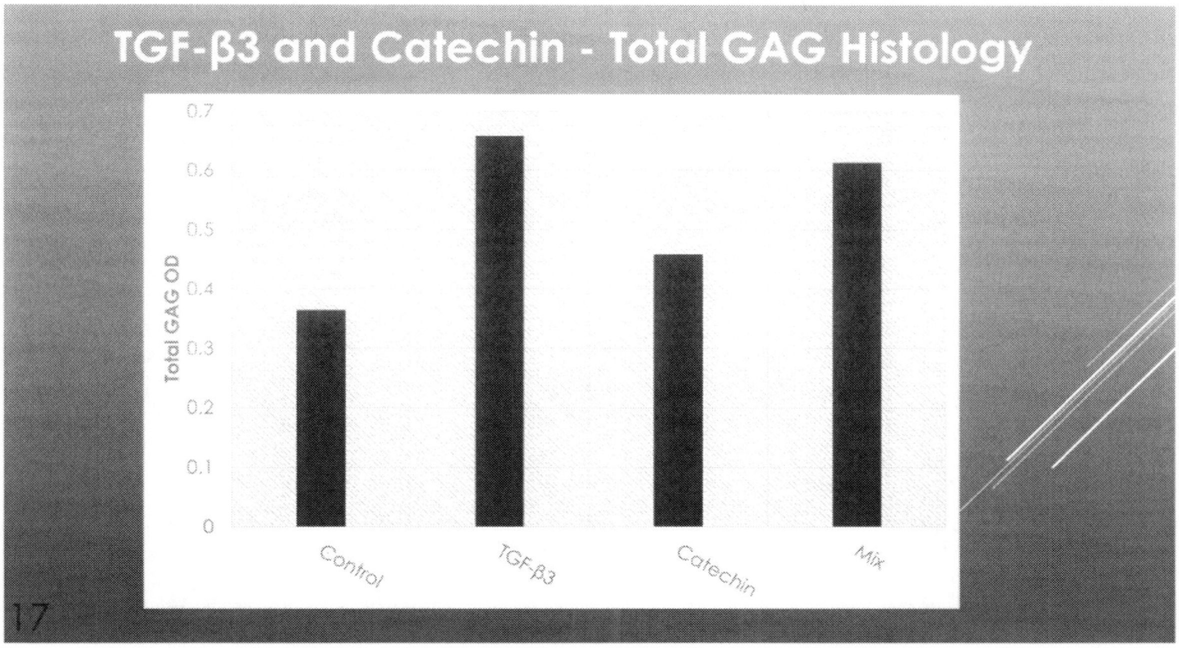
TGF-β3 and Catechin - Total GAG Histology
Total GAG OD
0.7
0.6
0.5
0.4
0.3
0.2
0.1
0
Control
TGF-β3
Catechin
Mix
17

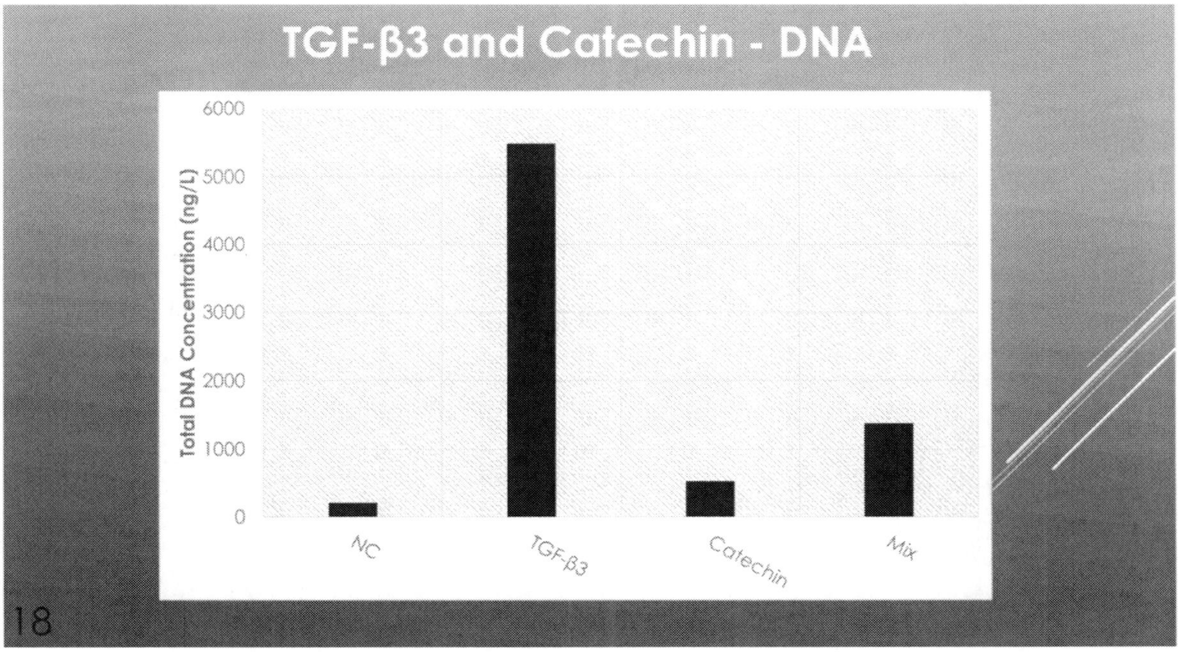
TGF-β3 and Catechin - DNA
Total DNA Concentration (ng/L)
6000
5000
4000
3000
2000
1000
0
NC
TGF-β3
Catechin
Mix
18

Future Direction

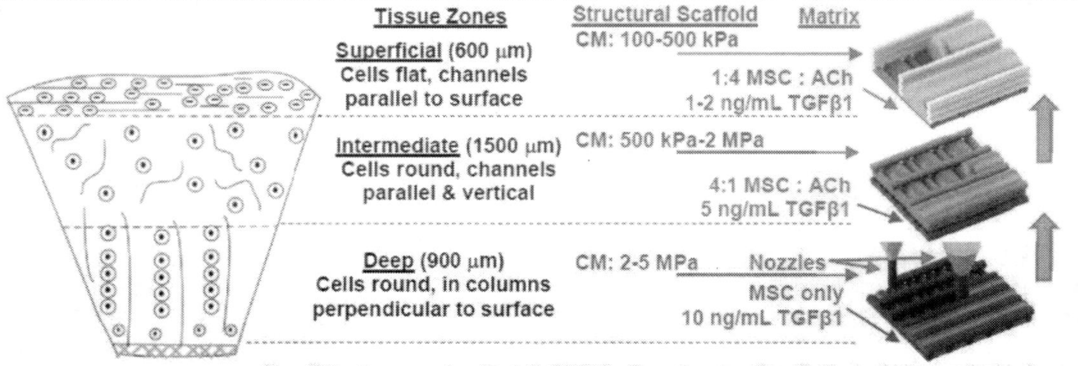

Key: CM – Compression Moduli; TGFβ1 – Transforming Growth Factor-beta 1, actual values to be determined in Objective 1

19

Acknowledgements

- Dr. Edwin Tingstad (Orthopaedic Surgeon)
- Dr. Vincent Idone (Collaborator at REGENERON)
- Haneen Abu Sharkh, Garry Smith, Somayeh Ramezanian, Samuel Uzoechi, and Baran Arsalan (Lab Mates)
- Bryce Delgado, Tyler Cheung, Jonathan Streat, Lujain Al Ali, Rachel Kagan, and Sarah Shroeder (Undergraduate Researchers)
- Funding NSF GOALI - 1606226

REGENERON NSF

20

THANK YOU!
ANY QUESTIONS?

496506 Quantum Molecular Sequencing: Unravelling Genomic Information One Molecule at a Time

Abstract:

Health and medicine are personal for each individual, but most of the current diagnosis and treatments follow â€œone size fits allâ€ approach. Recently there has been increasing awareness to remedy this, and several initiatives have been started towards personalized medicine, precision medicine, and others, to develop customized and tailored solutions. In this talk, I will address recent advances made in my group in developing Quantum Molecular Sequencing (QM-Seq) , a new nanoelectronic molecular spectroscopy technique that can provide a transformative solutions for detection, diagnosis and treatment of diseases. Particularly, my talk will focus on developing aÂ new paradigm in molecular recognition and nanoelectronic spectroscopy that will provide useful and relevant genomic information for development of personalized medicine and represent a unified platform for multiomics: genomic, transcriptomic, epigenomic, and RNA structure.

Nanoelectronic DNA sequencing can provide an important alternative to sequencing-by-synthesis by reducing sample preparation time, cost, and complexity as a high-throughput next-generation technique with accurate single-molecule identification. However, sample noise and signature overlap continue to prevent high-resolution and accurate sequencing results. Probing the molecular orbitals of chemically distinct DNA nucleobases offers a path for facile sequence identification, but molecular entropy (from nucleotide conformations) makes such identification difficult when relying only on the energies of lowest-unoccupied and highest-occupied molecular orbitals (LUMO and HOMO). In this talk I will present recent results on Quantum Molecular Sequencing, which utilizes nanoelectronic spectroscopy to develop a molecular recognition technique, using new biophysical parameters are developed to better characterize molecular orbitals of individual nucleobases, intended for single-molecule DNA sequencing using quantum tunneling of charges. For this analysis, theoretical models for quantum tunneling are combined with transition voltage spectroscopy to obtain measurable parameters unique to the molecule within an electronic junction. Scanning tunneling spectroscopy is then used to measure these nine biophysical parameters for DNA nucleotides, and a modified machine learning algorithm identified nucleobases. The new parameters significantly improve base calling over merely using LUMO and HOMO frontier orbital energies. Furthermore, high accuracies for identifying DNA nucleobases were observed at different pH conditions. Besides better characterization of molecular orbitals using the new biophysical parameters and specific biochemical conditions, Quantum Molecular Sequencing combines new algorithms and customized machine learning techniques for sequencing single-molecule DNA, RNA, epigenetic modifications, and RNA structure, using a simple comparison of QM-Seq electronic spectra with a library of respective signatures without any prior expectation. These results have significant implications for developing a robust and accurate high-throughput multiomics technique.

504791 Neuronal Biosensors

Abstract:

Biomimetics has stimulated innovation in developing medical devices, no less in the field of biosensing. The most exquisite sensors have already been created with, for example, crustacean antennae useful in detecting femtomolar (10^{-15} M) concentrations or less through neuronal chemoreceptors. Over the past 30 years our team has sought to understand and mimic neuron-like sensing. We began studying living neuron action potentials as a function of species concentrations. We turned to synthetic analogues isolating ion channels in bilayers and modifying structures to make responses sensitive to specific proteins. To further reduce these systems to practice we pivoted to inexpensive ionophore preparations in micro- and nanoscale ion selective electrodes. Finally, to make these systems practical for detecting cells and proteins we are developing a dual ionophore biosensing concept which ultimately mimics the neuronal system where we began. This plenary session will take the attendee from the fundamental physiology, through Nicolsky-Eisenman and Goldman-Hodgkin-Katz modeling, to synthetic practicality in the neuronal biosensor area, and finally to simple learning tools for teaching neuronal biomimicry.

505102 Mix (and Unmix) It Up with Biomembranes

Abstract:

Since the invention of the optical microscope in the 17th century it has been observed that biological membranes compartmentalize the cellular machinery of life. Then, nearly 100 years ago, it was concluded, through surface science, that the cell membrane is a lipid bilayer. Since that time, many observations and theories have been put forward to try to explain how complex behavior emerges in living systems from the ubiquitous lipid bilayer structure and its integrated proteins and carbohydrates. We are now in a period of time when some of the most important paradigms posited in the last several decades for emergence of complex behavior in living cell membranes are being tested and questioned. These include the membrane (or lipid) raft hypothesis and mechanisms for generation of membrane curvature and compositional asymmetry. An important outcome of this work may be the design of new micro- and nano-scale self-assembled systems and composites. This talk will focus upon our work in lipid demixing, crowding-induced mixing, sorting by nanoscale curvature, and bionanocomposites that may contribute toward the design of new biological membrane-inspired medicine and technology.

504790 Bridging the Gap Between Viral and and Nonviral Gene Vectors

Abstract:

Our understanding of disease, both genetic and acquired, continues to progress rapidly and provides insight into how most ailments can be treated at the genetic level. In spite of our improved understanding of genetic treatments, the field of gene therapy has failed to keep pace and advance human health. The lack of progress can be attributed primarily to the obstacles associated with delivery of genetic material. We been working to develop novel gene therapy vectors that bridge the gap between viral and nonviral vectors. We are working to combine the advantages associated with each respective vector while reducing or eliminating common drawbacks to gene delivery. We have primarily done so by replacing viral proteins with synthetic materials, such as polymers, lipids, or peptides. By working to understand how these materials interface with viruses and how they change the delivery pathway, we are aiming to gain knowledge necessary for designing novel materials that will help us progress step-wise from viral vectors to nonviral vectors.

486763 Protein and Gold Nanoparticle Based Radiation Sensor

Abstract:

High energy X-ray radiation has been used for ablation of malignant tumors for quite some time. Nevertheless, medical practitioners are unable to predict with high accuracy the amount of radiation reaching the point of interest below the tissue level. To help address this issue, we are developing a colorimetric sensor for radiation using the gold nanoparticle forming property of bio-molecules (particularly Maltose Binding Protein (MBP)) in presence of ionizing radiation. As gold nanoparticles have distinct color, gold nanoparticles formed on irradiation changes the color of the solution and thus can be used as colorimetric sensor. The color change is observed for doses of X-ray radiation as low as 2 grays (1 gray (Gy) = absorbance of 1 joule of radiation energy per kilogram of matter). The complete procedure for cloning and synthesizing MBP is also described. The system also shows a radiation dose dependent color shift (from 3 Gy to 15 Gy), which can be used to discriminate between different delivered doses.

498352 Direct Detection of Nucleic Acids without Amplification

Abstract:

Rapidly detecting the source of an infection remains challenging. Techniques using nucleic acid amplification have promise in providing answers quickly, but remain vulnerable to contamination and lack thermal stability because of their reliance on enzymatic proteins. We are developing a sensitive, inexpensive and portable instrument to automate the direct detection and identification of pathogenic nucleic acids, which will enable health professionals to rapidly identify infections in humans and subsequently accelerate the rational treatment of such infections. The current prototype performing the assay produces results from non-invasive samples in one hour and requires only minimally-processed crude samples. Our method promises to be robust for field and physician office applications, allowing distributed healthcare decisions outside of traditional laboratories.

All bacteria, viruses and fungi have within them nucleic acids (DNA, RNA) that determine their properties (benign or pathogenic) – and detecting the presence of specific nucleic acids is the fastest and the most optimal method to detect the source of an infection. Our methodology, called CAPTURE™ (Confirming Active Pathogens Through Unamplified RNA Expression), is unique in that it works in the presence of unknown amounts of unrelated nucleic acids, proteins and cellular debris. The design of the custom DNA 'capture' probes used in the assay is arguably the most integral step in our process, and as such, requires the most detailed method. Through a combination of locally compiled databases and proprietary software, ideal candidate sequences from pathogen genomes can be determined which exhibit the appropriate thermodynamic behavior required of the panel. The CAPTURE™ method is currently being used to identify several dozen specific pathogens and also to quickly determine the presence of any bacteria in a patient sample. The ability to determine universal bacterial signatures in a sample is a powerful tool for combating antimicrobial resistance, as it has the potential to limit the prescription of unnecessary antibiotics.

We have developed a CAPTURE™ panel for the analysis of urinary tract infections (UTI) that correctly identifies and differentiates the gram-negative bacteria *Escherichia coli* (which cause over 85% of UTI), *Klebsiella pneumoniae, Proteus mirabilis* and *Pseudomonas aeruginosa*; the gram-positive bacteria *Enterococcus faecalis* and *Staphylococcus aureus*; and fungal *Candida* species. The use of multiple captors for each pathogen allows for a statistical cluster analysis and introduces systematic redundancy to ensure efficacy. Our CAPTURE™ technique offers health care professionals a reliable and a quick way to screen for pathogens in clinical samples.

499935 PhotoacousticÂ Imaging to SimultaneouslyÂ Detect the Accumulation of Multiple Contrast Agents within Tumors

Abstract:

Modern cancer therapeutics have shifted focus from generic therapeutics to developing personalized treatment plans for greater safety and effectiveness. Since drugs are rapidly being developed specifically for each cancer phenotype, there is an increased demand for fast and inexpensive methods to determine cancer phenotypes and morphologies. Current diagnostic imaging modalities utilizing X-ray CT, MRI, and PET scans are limited to black-white images that cannot be used to differentiate multiple disease marker contrast agents at a time. In addition, the high costs of capital equipment and dangers of excessive imaging associated with these techniques inhibit their widespread use in diagnostics. Photoacoustic (PA) imaging, a hybrid light and sounds imaging technique, has shown to be a safe and inexpensive diagnostic technique with high spatial resolution in 3D. Traditional PA contrast agents, however, tend to have broad absorption peaks in the NIR range which renders it difficult to simultaneously image more than one signal at a time in deep tissue. Here we present the formulation of a series of PA active NPs with sharp and separable absorbance profiles in the NIR range for simultaneous multiplexed imaging (**Fig. 1**). PA dyes are encapsulated inside NPs using the controlled self-assembly mechanism, Flash Nanoprecipitation (FNP). Four new contrast agents, with sharp absorbance maxima between 600-900 nm, were created by encapsulating a variety of phthalocyanine derivatives. We were able to concurrently detect the concentrations of contrast agents mixed together with >95% deconvolution efficacy. Using these dye particles, we demonstrated that information on tumor morphology and surrounding blood flow can be acquired simultaneously (**Fig. 2**). As proof of concept, we co-injected RGD modified NPs and non-modified NPs with different labeling agents and tracked NP biodistributions for both particles simultaneously. Using this technology, we accessed the effect of NP ligand modification on both targeting efficacy onto the tumors and off-targeting accumulation in the liver using a single animal model. Over modification of the NPs resulted in rapid liver clearance and poor accumulation in the tumor; at low modifications, the tumor to liver accumulation ratio is 9.9 ± 4.2, while at high RGD modifications the tumor to liver accumulation ratio is 52 ± 22. Our particles enable new quantitative and rapid methods of assessing nanoparticle biodistributions.

500760 DNA Methylation Detection with an Engineered Protein That Binds Hemi-Methylated DNA

Abstract:

DNA methylation, specifically the methylation of cytosine bases, is an important biomarker, as abnormal DNA methylation patterns are found in many different types of cancer. Currently, a small number of cancer hospitals evaluate the methylation status of the MGMT gene promoter to determine the best course of treatment for patients with glioblastoma. However, improved methylation detection techniques are required in order to expand the availability of such testing to more patients. Methyl-CpG-binding domain (MBD) proteins bind specifically to methylated DNA sequences, and many assays have been developed that use these proteins in methylation profiling of DNA. The wild-type proteins in the MBD family bind specifically to symmetrically methylated CpG dinucleotides. Here, we describe the engineering of a new MBD that binds to hemi-methylated DNA but not unmethylated DNA, which allows for the detection of a methylated target sequence hybridized to an unmethylated DNA probe. The binding affinity of this engineered protein toward a DNA sequence with a single hemi-methylated CpG dinucleotide was evaluated, and a dissociation constant of 5.6 ± 1.4 nM was achieved. The protein was expressed in soluble form as a fusion protein with green fluorescent protein, and performance in distinguishing between hemi-methylated (methylated target DNA bound to an unmethylated probe) and unmethylated (unmethylated target DNA bound to an unmethylated probe) DNA in an interfacial binding assay was demonstrated. Using this protein, an assay was developed and optimized to determine the methylation status of the MGMT gene promoter by hybridizing the target DNA to a probe on a surface and determining the methylation status of the captured target based on the amount of MBD bound. Such an assay is useful in guiding treatment plans for glioblastoma patients because MGMT promoter methylation is correlated with better response to alkylating chemotherapeutics such as temozolomide.

501477 Effective Physiological and Anatomical Parameters on Fractional Fluid Reserve (FFR) of Coronary Artery Stenosis

Abstract:

Coronary stenosis is an abnormal narrowing in a coronary artery that is caused by an atherosclerotic lesion and reduces lumen space. FFR is a gold standard method to assess the hemodynamic significance of coronary stenosis. FFR is the ratio of maximum blood flow distal to a stenotic lesion to normal maximum flow in the same artery. Normal FFR is 1 and an FFR < 0.8 is considered hemodynamically significant. FFR is typically measured invasively in the cardiac catheterization laboratory using a pressure sensor wire in the coronary artery. Recently, investigators have applied computational fluid dynamics (CFD) methods to determine FFR using coronary images obtained from different modalities such as angiograms or CT scans. The purpose of this study was to determine the effect of eleven physiological and anatomical factors on the accuracy of computing FFR. In this study, blood flow was modeled as a single phase Newtonian fluid. The physiological factors were pressure as inlet boundary condition, mass flow rate as outlet boundary condition, heart rate, blood viscosity and blood density. Anatomical factors were the diameter of coronary artery, length of coronary segment modeled, percent stenosis, stenosis length, the position of stenosis relative to coronary segment modeled, and stenosis shape (concentric vs. eccentric stenoses). The parameters that affected FFR the most were determined by Plackett–Burman statistical analysis. Analysis of variance (ANOVA) showed a high variance coefficient (R^2) value of 0.928. Our analysis demonstrated that percent stenosis, diameter of coronary artery, mass flow rate, and stenosis shape had the most significant effect on FFR ($p < 0.05$). Increase in mass flow rate decreased FFR significantly, while the diameter of the coronary artery was directly proportional to FFR. As for stenosis shape, FFR decreased more significantly due to concentric stenosis than eccentric stenosis. CFD provided analysis of coronary physiology that would have been difficult or impractical to do experimentally. Accurate anatomical reconstruction is critical for computationally deriving FFR.

Keywords: Fractional Fluid Reserve; Coronary Artery Stenosis; Plackett–Burman Method; Computational Fluid Dynamics;

501819 Analysis of Carbonyl Compounds in Exhaled Breath for Identificationf of LUNG Cancer Biomarkers

Abstract:

Some carbonyl compounds have been identified as biomarkers of lung cancer. We report a fabricated reactive chip with thousands of micropillars in a microfluidic channel that have been engineered to selectively preconcentrate trace volatile ketones and aldehydes in exhaled breath. These carbonyl compounds are directly related to lung cancer induced metabolic dysfunction and oxidative stress. The chips are fabricated using deep reactive ion etching (DRIE) of silicon. The micropillar surface is functionalized with a reactive quaternary ammonium aminooxy salt for trapping trace carbonyl compounds by means of an oximation reaction. The chips are used for directly preconcentrating carbonyl compounds in breath. The concentrated ketone- and aldehyde-aminooxy adducts are analyzed by ultra-high performance liquid-chromatograh-mass spectrometry (MS). Some carbonyl compounds including hydroxyl-acetaldehyde, 2-butanone, 4-hydroxy-hexenal are identified as lung cancer biomarkers by statistical data analysis of healthy control group breath samples and lung cancer patient breath samples.

AUTHOR INDEX

AUTHOR INDEX